594.6
HAM

22.3.73

POLLUTION: the world crisis

Tom Stacey Ltd., 28-29 Maiden Lane,
London WC2E 7JP, England

First published 1970
4th Impression 1972

© Lynette Hamblin 1970

ISBN 0 85468 001 2

Printed in Great Britain by
Biddles Ltd., Guildford, Surrey

POLLUTION:
THE WORLD CRISIS

Lynette Hamblin

TOM STACEY LTD

CONTENTS

INTRODUCTION: Spaceship Earth		vi
I	Upsetting the natural balance	1
II	Dead lakes, dying rivers	11
III	The sea, sewer to the world	31
IV	Killer oil	44
V	Pesticides, the indiscriminate killers	62
VI	Atmosphere in peril	85
VII	Nuclear Pollution	102
VIII	Too many tame people	117
IX	Too few wild animals	125
X	Engineering havoc	141
XI	How the world must act	150
GLOSSARY		162
INDEX		164

(handwritten margin note: "DDT a threat →" pointing to chapter V)

Author's note

Most of the information in this book was obtained from the following publications: **Nature, Science, Environment, New Scientist, Scientific American, The New York Times**, and the press reports of the World Wildlife Fund. The following were also of assistance to the author: **The Unesco Courier** of January 1969 and March 1969, the report of the Secretary-General of the UN Economic and Social Council **Problems of the Human Environment**, the BBC Reith lectures by Dr. Frank Fraser Darling, **Silent Spring** by Rachel Carson, **Pesticides and Pollution** by Dr. Kenneth Mellanby.

INTRODUCTION

Spaceship Earth

Buckminster Fuller's analogy with a spaceship is exact: a small, vulnerable, delicately balanced mechanism that is above all finite. Nothing we have is indestructible or inexhaustible — not the air we breathe, not the land we use, not the water we drink; they all have an end, increasingly a foreseeable end. The strongest impression I got from the photographs brought back from space was of a very small, very lonely planet. And so it is, and so it must be treated.

In this book we will see how tenuous our hold over the world is and in how many ways we are misusing it. At the moment it is an open question whether Man will destroy his only home by blowing it to pieces with the hydrogen bomb or, less quickly but equally surely, making it uninhabitable by pollution. We will examine all the various ways — obvious and not so obvious — that we are putting our future at risk.

Those same space photographs show a blue-green planet of rare beauty that contains all the ingredients for life. A fertile land, a breathable atmosphere, plenty of water and vast natural resources. But a closer look at what Man has done with these ingredients reveals a different picture: the earth scarred and torn in the effort to extract precious minerals; a thick pall of polluted air lying heavy over many cities; rivers and lakes choking with sewage and rubbish; the sea poisoned by chemicals and oil; and the dominant species proving his superiority by destroying the other animal species on which he depends. A dismal list of ignorant opportunism and thoughtless arrogance.

We have now reached a point in time when, however sketchy our knowledge, we can with some accuracy predict the results of our actions and see what decisions we must make about priorities.

<u>Pollution is clearly bound up with population. Too many people mean too great a demand on the world's resources</u>. In effect, we are eating up the world — land, sea and air — too fast. <u>The struggle to provide for the population in food and industrial goods has contaminated the whole biosphere</u>. Equally, too many people mean too much waste and waste, moreover, that is not easily destroyed — toxic minerals, persistant chemicals and radioactive materials.

In the past there were not enough people — whatever they did — to affect adversely more than a small part of the world around them. It could always recover. Today the attack is on too broad a front, too cumulative and irreversible, to permit Man's traditional attitude that the natural resources of the earth are infinite. The natural resources may be very great, but they are not limitless. Similarly, the ability of nature to repair the damage that Man does is great but it is being exploited and abused. If he continues to exploit the earth at his present rate, he will render it unfit for human habitation within a few generations.

Time and again we will see that Man is not a being that can transcend the world about him. He is indissolubly a part of it, one cog in a giant mechanism. The significance of the population explosion, then, goes much deeper than just a shortage of food. It entails a total overloading of the balanced system in which he lives.

<u>Man cannot be allowed to continue increasing indefinitely.</u> If the population goes on rising at the present rate there will be 25 thousand million people on earth in 100 years time — 1,000 people per square mile of the earth's habitable area. It cannot support this burden. If the population growth is not to be curbed voluntarily, and assuming massive wars of destruction are avoided, starvation and disease will do the job for us. Already we are sacrificing the quality of human life in the interests of the quantity of humans.

This is the only home we have. If we wish to survive we must learn to treat it more wisely. Wholesale pollution of the land, sea and air, exhaustion of the mineral resources under the earth, and elimination of the plant and animal life on the surface, will result in the destruction of Man's only habitat. If the infinitely complex systems that make life possible are damaged, the earth will stop working and all life will die away.

I

UPSETTING THE NATURAL BALANCE

Life is found from the ocean floor to the peaks of the highest mountains. Life can survive in the bitter cold of the poles and in the scorching heats of the waterless deserts. But most of it is concentrated in more hospitable regions and in the upper 500 feet of the oceans. What makes this life possible is the dynamic equilibrium between the atmosphere, the soil and the plant and animal life on land and sea. No factor is fixed for ever; and no part of the process is not necessary to the existence of the whole. The entire realm of the biosphere — the narrow layer of water, land and air where life exists — is the result of continuous actions and reactions. To change one element is to risk changing the whole structure. In the world around us nothing is immutable.

The oxygen in the air we breathe is a case in point. The atmosphere of the primitive earth when it was first formed contained no free oxygen at all: originally it was probably high in hydrogen, water vapour, methane and ammonia. The present atmosphere, about 78 per cent nitrogen, 21 per cent oxygen and very small amounts of argon, neon, carbon dioxide and a number of other gases, is a product of all the life in the biosphere and is maintained by the plants, animals and bacteria. Primitive organisms living in the original atmosphere

POLLUTION: THE WORLD CRISIS

produced minute quantities of oxygen and the long road to the creation of the atmosphere as we know it was begun.

Oxygen in the air is constantly being consumed by the animal life on earth which uses it and then breathes out carbon dioxide. If the oxygen was not replaced and the carbon dioxide removed, animal life would eventually die. The level that is essential for life to survive is maintained by the balance and efficient functioning of two basic, closed cycles — the carbon and the nitrogen.

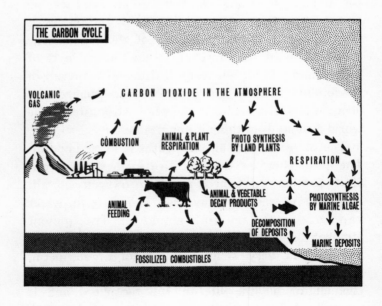

UPSETTING THE NATURAL BALANCE

There are three clear and separate stages in the carbon cycle. The primary step is the conversion of carbon dioxide from the air into glucose and oxygen. This takes place within plants by the process called photosynthesis. Chlorophyll, besides giving the plant its green colour, is the catalyst that takes up the sun's energy and allows a number of complex chemical reactions to occur during which the carbon dioxide from the air and water from the soil are converted into glucose and oxygen, which is released into the air. The glucose can then be converted into other useful carbon compounds within the plant (carbohydrates etc.). The release later of the energy stored in some of these compounds, by oxidization, gives off carbon dioxide which is returned to the air. If the plant is uneaten the carbon compounds return to the soil when the plant dies and rots. Photosynthesis thus combines carbon dioxide, water and the sun's energy and creates and sustains life.

More than that, plants are the ultimate source of food for all animal life. For though an animal may itself be carniverous, yet at some point it must prey on herbivores — the start of every food chain is some form of plant life.

The next step in the carbon cycle is the animal eating the plant. Within the animal the carbon compounds are either oxidized to produce energy and carbon dioxide is breathed out or they are stored in the animal tissues and return to the soil when the animal itself dies. Some are excreted and return to the soil in that way.

POLLUTION: THE WORLD CRISIS

The decomposition of animal and plant tissue by bacteria and fungi in the soil and water forms the third part of the carbon cycle. The proteins, carbohydrates and fats are all broken down, oxidized and returned to the air as carbon dioxide which the living plants then use to build up their tissues. So the cycle endlessly continues.

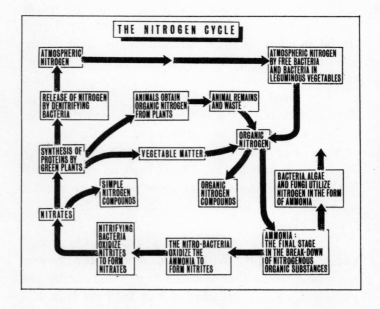

UPSETTING THE NATURAL BALANCE

The other essential cycle is the nitrogen. Nitrogen is found in all living beings as proteins and amino acids. Again, it starts with plants. They absorb nitrogen in the form of nitrates from the soil; and incorporate it into complex carbon-based molecules — amino acids — which are linked up to form proteins. Animals then eat the plants, in turn incorporating the proteins within their bodies. When the plants and animals die, the proteins return to the soil where they are digested by the micro-organisms that reconvert them into the nitrates and ammonium salts that the plants can use. Some soil bacteria can convert or 'fix' the inert nitrogen in the atmosphere into their cell structure and thus, when they die, make it available to help build up the stores in the earth and maintain a healthy land. These are found in the roots and nodules of leguminous plants like clover, peas and beans.

Within the biosphere there are many different types of ecosystems but the keynote in all of them is balance. Balance between the prey-predator populations in animal species, balance between the plant and animal life in one area, balance between the plants and the land. A clear example of just such a crucial balance is shown by the work of U.S. scientists taking part in the International Biological Programme who are studying, among other ecosystems, the grasslands of the American West, a major project extending from the Rockies to the Mississippi. They are examining all possible interrelationships in this habitat in order to find out what changes Man can safely make to the environment to increase food supplies. One fascinating interrelationship they have already found is that between the cow and the lark bunting, which reveals the simplicity and logic of natural balance. Cows eat a plant, the saltbush, that the lark bunting lives in. As the cows eat the available saltbush, the lark bunting population naturally decreases. On the face of it this would seem to be of little significance, but unfortunately one of the bird's normal functions is to help keep the grasshopper and other insect population in check. So when the lark bunting is eliminated, the grasshoppers increase and begin to compete with the cows for the grass. Eventually if this process is allowed to continue, a plague of insects can develop and pose a serious threat to the cows' survival.

Lack of balance is causing the destruction of the Great Barrier Reef, off the Australian coast, and threatening the coral reefs of the Pacific islands. The proliferation of the crown of thorns starfish has already killed over 140 miles of the Great Barrier Reef and the 'epidemic' has spread to Guam, Palau, Saipan, Rota and other Pacific islands. The starfish feeds on the living coral (which not only provides a rich source of food for fish but also protects the coastlines from erosion), digesting the tiny polyps and leaving a dead white skeleton behind which is easily washed away. Once the crown of thorns was comparatively rare but since 1963 there has been a population explosion (the female lays between 12 and 14 million eggs in December and January) and now herds of starfish 200-strong destroy over half a mile of coral reef every month. The reasons for this catastrophic increase are not clear; perhaps the depletion by shell collectors of its natural predators, the giant triton snail, helmet shells and clams, is partially responsible. Other possibilities include the dredging and blasting operations on the reefs, which may have provided breeding grounds where the young starfish are safe from attack by the coral which eats the larval stages. DDT in the Pacific Ocean may be playing a part. Intensive research into the problem is being carried out by Dr Richard Chesher and his colleagues at the University of Guam.

The study of ecology has become vital not only on a local but on a world-wide scale. Man today has the means to destroy the whole biosphere. In the past, when Man's numbers were limited, his ability to damage the environment was limited. One small tribe cutting down and burning vegetation and moving on to another site when the soil was exhausted had little effect on the environment. It could recover when they moved on. Nature's capacity for self repair was not overtaxed. Today, two-thirds of the world's forest area has been destroyed. Since forests help maintain our supplies of oxygen and remove carbon dioxide from the air, it is no longer only local ecosystems that are affected but the whole biosphere.

Burning fossil fuels (coal, oil and gas) has increased the carbon dioxide content of the atmosphere by 15 per cent in the past 100 years. The consequences of this are complex. Normally the oceans and vegetation remove carbon dioxide from the air. However the increase in carbon dioxide led — until very recently — to a warming up of the earth's atmosphere. This increased the capacity of the land to remove carbon dioxide by encouraging tree growth (though Man then cuts them down), but it decreased the sea's capacity, since warmer water takes up less carbon dioxide. (The net result of this and counterbalancing processes are discussed in Chapter VI.) It could lead to the melting of the icecaps and inundation of the low-lying areas of the earth. Hence Man's technology is attacking the natural balance from both angles; by pouring excess carbon dioxide into the atmosphere and at the same time impairing the natural processes that remove it.

Perhaps the most frightening way Man could destroy the biosphere is by the effect of modern technology on the oxygen supply. It has been seen that the oxygen we breathe is the product of the life around us. Inhibit or destroy that and Man will suffocate. Approximately 30 per cent of the oxygen supply is produced by phytoplankton or marine plants and animals. The pesticides currently used which are found all round the world can poison those plankton. Add to this the wholesale destruction of the forests and the picture is bleak indeed.

Man has behaved like an impetuous child. He has established dominion over all the life on earth only to destroy it. Unfortunately he will destroy himself as

UPSETTING THE NATURAL BALANCE

well. The treasures of the earth have been regarded as infinite. At this eleventh hour Man is beginning to realize this is not so.

We will see throughout this book that every part of the biosphere is being threatened by pollution and misuse. Rivers and lakes are endangered by the sewage and industrial waste being poured into them, rendering them unfit to drink or for fish to live in. The excess of nutrients allow algae to flourish, but the subsequent deoxygenation of the water suffocates the fish and turns lakes and rivers into stagnant cesspools. The land is becoming impoverished by intensive cultivation which removes essential minerals. Even artificial fertilizers can have a deleterious effect on the soil. The sea, potentially the major future source of food, is being poisoned by chemical pollution and many species of fish are being reduced or wiped out by over-fishing.

The air we breathe is laden with smoke and dust and lethal chemicals. Finally we are squandering the earth's resources at an alarming pace. The fossil fuel reserves, at the present rate of exploitation, will be exhausted in the next 100 years or so.

There is no corner of the biosphere which we are not harming — perhaps permanently — and no corner of the biosphere can be harmed without the rest being put at risk. If Man is to survive, he must radically change his attitude towards his environment. We may be the dominant species but we are still dependent on all the other forms of life on earth for our survival. At the present rate — although patchy

POLLUTION: THE WORLD CRISIS

and sporadic attempts are being made to combat pollution — Mankind is hell-bent on self-destruction.

II

DEAD LAKES, DYING RIVERS

Filthy, brown, stinking rivers clogged with heavy growths of algae and foaming with detergent wastes. Vast sheets of lifeless, stagnant water where once there was a lake teeming with living things. A black, heaving, viscous scum of oil. Fish floating white belly up. Banks littered with rusty tins, broken bottles, all the sodden refuse of our throw-away society. A world choking and drowning in its own faeces. It is a nightmare and yet one that is now commonplace all over the world. For too long every sort of waste has been dumped into the rivers and lakes. As far as the offenders are concerned — out of sight, out of mind. But here the wages of sin are death — death of the water on which our civilisation depends no less than every civilisation in history. And what is worse is that with increasing population the problem increases.

Many rivers are now so badly polluted that their water is unusable for drinking purposes. Industrial waste, miscellaneous chemicals, raw or poorly treated municipal sewage, nitrates from fertilizers and hot water from power stations are all discharged into the rivers and lakes. Water has a certain capacity for self-purification but it is just not capable of cleaning the overwhelming flood of effluents that modern society pours into it. We are literally swamping our rivers and lakes.

POLLUTION: THE WORLD CRISIS

The mild-sounding scientific name given to the process of the 'ageing' of lakes and rivers, caused by overfeeding of the water, is eutrophication. It is being reported from all over the world. As a river flows further away from its source, more and more nutrients from the soil are leached into it in drainage. Added to this are the large quantities of nutrients from sewage plants. This encourages plant growth in the rivers and lakes leading to the deoxygenation of the water — the more nutrients, the more plants, the less oxygen and therefore the older the waters become.

In nature eutrophication is a slow process. Plant growth can normally be held in check by the animals that live on them and a balance is struck. However, the vast quantities of nutrients that are being poured into the rivers and lakes in fertilizer run-off from agricultural lands, and in sewage from towns, are grossly overloading the system. This very high level of nutrients, especially nitrates, in the water encourage an excessive growth of algae which reduce the oxygen content of the water. When the oxygen content falls below a certain level, the aquatic life slowly suffocates leaving no competition for the plants. They grow fast and then die and rot, leaving stagnant smelling water even more inimical to animal life. The lake dies.

Lake Erie exemplifies this process. The combination of nitrates from agriculture and sewage and phosphates from detergents have prematurely aged the lake by some 15,000 years. Each year 90

DEAD LAKES, DYING RIVERS

million pounds of nitrogen from municipal sewage and 75 million pounds from fertilizers are added to the lake, and it would take an estimated $40 billion to clean it up. There are now hardly any fish left. The condition of the other Great Lakes, though not so badly polluted, is nevertheless grave. Because the lakes have a surface area of 95,170 square miles, they have been treated as if they had an infinite capacity to absorb waste matter of all kinds. The cities of both the U.S. and Canada have poured their industrial wastes and municipal sewage, in many cases untreated, into the lakes with abandon for many years. Predictably, the water quality has been severely impaired and its use for domestic and industrial purposes endangered.

The whole area's highly profitable tourist industry is threatened by the increasing pollution. A number of beaches have been declared unfit for bathing and the game fishing in the area has been greatly reduced by pesticide poisoning. The increase in the population of the Great Lakes area has meant an inevitable increase in sewage with a consequent need for more purification plants. Vast sums of money will have to be spent to avoid polluting the waters any further. Already poorly treated human wastes, assisted by animal wastes, have been responsible for increasing the bacteria contents of Lake Michigan and Lake Superior and that of the coliform bacteria has reached a dangerous level for human consumption. There needs to be much stricter enforcement of the existing water quality control regulations.

POLLUTION: THE WORLD CRISIS

In addition to the enormous amounts of normal human and industrial wastes poured into the lakes from the vast industrial complexes on their shores, there is a growing problem of accidental pollution disasters. In 1969 10,000 gallons of soya bean oil flowed into Lake Michigan when some equipment failed at Proctor and Gamble's Chicago plant. This sort of accident may be unavoidable: what is not unavoidable is that so little is done to minimise or isolate the effects of such accidents.

Industrial and chemical pollution, oil wastes from shipping, pesticides and thermal pollution all threaten the Lakes. The problem is an international one — Canada and the U.S. both use and pollute the waters. A number of States and Provinces in both countries must co-operate if anything is to be achieved. Unilateral action by individual towns will not materially affect the situation. So far this vital national and international co-operation has been elusive.

The Canadian plan for drilling for gas beneath Lake Erie is currently causing concern in the U.S. Canada owns half the lake and wishes to drill for the natural gas which is needed in the Ontario region where energy is in short supply. The U.S., owning the other half, fears that pollution will follow such a move, and Lake Erie's condition will become finally irremediable. One possibility is that oil will leak from the wells — this happened in 1959 when oil blew out of a well being drilled in western Lake Erie. Another possibility is that brine could leak from the well, which could prove disastrous for the remaining life in this 'fresh-water' lake. The U.S. halted drilling in the Lake in 1969. It is, however, unlikely that the Canadians will follow suit. Thus the Great Lakes, an area of great natural beauty and a favourite recreational area for Americans and Canadians alike, are slowly dying and there seems little reason to expect they will not continue to do so.

One major contributor to Lake Erie's state is the River Cuyahoga. It has the dubious distinction of being the only stretch of water in the world to be classified as a fire hazard. In July 1969 the oil on the surface combined with gases from the sewage effluent in the river bed, ignited and nearly burned down two bridges crossing the river. The river itself is completely dead. Federal Water Pollution Control Administration noted 'the lower Cuyhoga has no visible life, not even low forms such as leeches and worms that normally thrive on wastes.'

Further south another American lake — Lake Apopka in Florida — is as badly polluted as the Great Lakes. Since the mid-1940s it has become so polluted that only rough fish, such as the gar, have been able to survive. The lake has been ruined by sewage, insecticides and industrial waste from the local citrus industry. Much rich soil has been washed into the lake, which a biologist described as being 'too thick to navigate and too thin to cultivate'. It covers 50 square miles. In an attempt to rid it of the pollution, draining began in February 1970, and will continue until December, when it will be halted during the citrus season. (In cold weather the land surrounding the lake, where the plantations are, is kept warm by the heat radiating from the water and so the fruit develops faster than usual.) Once the citrus season is over, much of the remaining water will be pumped out at the rate of 750,000 gallons a minute. This should expose 85 per cent of the lake's bottom to the sunlight. It will be baked in the sun for several weeks until the assorted mud and pollutants harden. Then it will be refilled from a giant freshwater spring. If successful, this $140,000 experiment will be repeated in a number of other Florida lakes. Unfortunately, it is obviously not feasible to treat the Great Lakes in the same way.

For some time the Swedes also have been experimenting with techniques designed to regenerate dead or dying lakes. Scientists from the Limnological Institution at Lund University are working on an eleven-year project to bring Lake Trumman, near the centre of Vaxjo, back to life. It was killed by a combination of raw sewage and industrial waste. They are pumping out the top 20 inches of the 16-foot thick blanket of dead matter that coats the bottom of the lake. This layer contains most of the pollutants and its removal should have a revivifying effect. The water of other lakes is being reoxygenated by pumping the dead bottom water up and running through artificial streams to aerate it and remove poisonous hydrogen sulphide: it is then returned to the bottom. Water has also been reoxygenated by pumping air into the bottom of lakes. This also causes the water to rise to the surface where it picks up further oxygen. There are risks in this method. The air has to be pumped in carefully. If the dead water rises too fast, it can suffocate the life in the upper levels of the lake. All these approaches are still in the experimental stage and the results will not be available for three or four years.

A method which has been applied to lakes with a high phosphate content is to spray them with aluminium sulphate. This reacts with the phosphorus to form a heavy gel which sinks to the bottom of the lake forming a harmless blanket. Two lakes near Stockholm were treated by this method and have shown a marked improvement. Algae formation has nearly dropped to zero and the lakes are much clearer, but how lasting this will be remains to be seen.

Eutrophication is a world-wide problem. Lake Baikal in the USSR is already showing the first signs. It is the world's deepest lake and an enormous freshwater reservoir but it is being threatened by industrial development on its shores. This situation is only one of many being closely watched. The Russians are already concerned by increasing pollution and interference in the ecological balance of many lakes and rivers. One significant sign is a drastic fall in the catch of sturgeon, thus endangering the caviar industry. The reason is that the sturgeon's natural habitat in the Caspian Sea has been shrinking. Due to diversion of river water into irrigation

POLLUTION: THE WORLD CRISIS

schemes, the level has fallen by more than five feet in the past few decades and its area reduced by thousands of square miles. In addition to this, the sturgeon have been decimated by oil pollution and industrial waste both in the Caspian itself and in the rivers by which it migrates to its mating grounds. So few have been managing to survive the journey up these rivers that they are now being loaded onto river boats which carry them safely over the migration routes. The Volga, which discharges its filthy water into the Caspian, carries the waste products of some of Russia's biggest factories, in addition to the human wastes of large towns. The Ural, which also runs into the Caspian, is likewise very badly polluted. Added to this, the Ural is getting so shallow, due to its water being drained off for irrigation and industrial use, that the water supply to towns is being affected. The sturgeon catch has been reduced to a fraction of its former value and it is nearly impossible to buy caviar in Russia except in restaurants and for export.

Here, as in all the industrial areas of the world, there is need for a greater number of water purification plants, sewage farms and strict control of industrial effluent. It is essential to have vigorously enforced legislation to ensure a balanced use of the environment.

The position in Europe is not that much better. Only a few diseased eels now manage to survive in the lower reaches of the Seine, and the Rhine is regarded as the open sewer of Europe. The Rhine made news in June 1969 when a couple of barrels of the insecticide endosulphan were accidentally tipped into it. Some 40 million fish were killed and the Dutch, who have the misfortune to live at the wrong end of the river, were forced to turn to emergency water supplies.

If the Rhine is the sewer of Europe, Holland is surely its cesspool. They are the perennial sufferers from the normal pollution of the river. Swiss, French and German industries all contribute to the pollution but the decisions of the only international organisation, the International Committee for the Protection of the Rhine River against Pollution, are not binding on its members. In the case of the insecticide disaster in June 1969, there were two vital days wasted before the Germans passed on any information about the leakage. Compensation in the event of a disaster is practically impossible to collect. Hence Holland is dependent on voluntary action by the other states. In November 1969 the condition of the river was such that Rotterdam was forced to stop using the water and switch to ground water reserves which are very limited. The situation was aggravated by weeks of inadequate rainfall which meant that the Rhine current was not strong enough to force the pollutants out into the North Sea. Here again is an instance of the necessity of international co-operation and here again we see how great are the pressures of finance, national prestige, and vested interest against it.

That it *is* possible to clean grossly polluted rivers has been shown by Britain's experience with the

Thames. It is rapidly coming back to life. For nearly a century there has been little or no fish-life in the lower reaches of the Thames, but during the past ten years fish have started to reappear. In September 1969 anglers caught 161 fish of 41 different species in the Thames below Fulham Power Station. The fish had fought their way up the Thames past the blackest patches of pollution and still survived. Only a few years ago the stretch between Gravesend and Putney was so polluted that it was an impassable barrier to fish. The improvement has been due to the strict powers granted to the Port of London Authority to penalize polluters, coupled with improvements in sewage purification techniques by the Greater London Council — formerly one of the worst offenders. The GLC hope to complete a giant £20 million purification scheme by 1973 which should further clear the water.

Strict enforcement of the 1951, 1960 and 1961 anti-pollution Acts in Britain has done much to improve the quality of water, although much remains to be done. Of the total length of the rivers in Britain over 5,000 miles are grossly polluted. Most of the worst cases occur in the heavily industrialised regions.

One of the main difficulties is that enforcing the pollution legislation, even when it exists, is left to the local councils. They have most to lose if, by insisting on effective and often costly anti-pollution measures, local industry is forced to close up and go elsewhere. The prosperity of the area suffers and there is little political advantage to be gained by helping to raise unemployment figures.

Water has a natural capacity for self-purification. Any organic matter, like sewage, is broken down by aerobic bacteria which require oxygen to work. Initially their population increases to counter the increased pollution but this leads to a decrease in the available oxygen and the purifying bacteria eventually stop working. At this stage anaerobic bacteria, which do not need free oxygen, start operating and these are responsible for both the putrefaction of plant and animal life and also produce noxious smelling compounds such as hydrogen sulphide. The oxygen level is sustained by the balance of organisms present, some consuming, others producing oxygen, and by supplies of fresh clean water. It is where this fresh water is taken directly from the source of the river for human consumption that most danger occurs. Where vast quantities of sewage are flushed into the river the biological self-purification system is overloaded and the river becomes deoxygenated. Aquatic life dies and the water, lacking anything to renew it, becomes foul and stinking. Dr Kenneth Mellanby, director of the Monks Wood Research station, Huntingdon, showed in his book *Pesticides and Pollution* (1967), that it would only take the raw sewage from 100,000 people completely to deoxygenate the River Thames (flow-rate, 200 million gallons per day).

Most sewage plants do only two things: purify the

water by removing the solid matter, and then break down the organic matter left using bacteria. The 'clean' water that they release into the rivers, however, is not so clean. It has a high level of nutrients, phosphates and nitrates, and these, as we have seen, encourage the growth of algae. This kind of pollution by heavy growths of algae, encouraged by inadequate or non-existent sewage plants, has been recorded in many reservoirs and rivers the world over. When the river has to cope with nitrates from fertilizer run-off as well, it is overloaded and gives up.

Of further concern is the hazard to humans of the increased level of nitrates in the water supply. The nitrate itself is comparatively harmless but certain intestinal bacteria convert it into nitrite which is poisonous. It combines with haemoglobin in the blood to form methaemoglobin which cannot transport oxygen. As a result, the body suffocates. Children under three months are particularly susceptible to this form of poisoning and a number of deaths in Europe and America have occured

Many sewage plants are thus totally inadequate. President Nixon has a $10 billion plan to build new plants: but only primary and secondary municipal water-treatment plants. The primary plants remove the solids in sedimentation tanks; the secondary, the organic matter through biological treatment by bacteria. It is the more expensive tertiary plants which are vital because they kill both infectious bacteria which survive the other treatments and remove the nutrients that cause eutrophication.

This part of the sewage process is expensive because it consists of letting the water stand in large artificial ponds for seven days and this uses a tremendous amount of land. It is through oxidation that the bacteria and nutrients are removed. After being passed through the ponds the water is approximately as clean as that which the water boards take out of rivers to purify for drinking purposes.

The American Chemical Society's report, *Cleaning Our Environment. The Chemical Basis for Action*, recommended that radically new solutions to the treatment of sewage based on microbiological techniques should be sought. The ultimate, of course, is the direct conversion of polluted water into drinking water. A prototype is working in Lebanon, Ohio. The cost of such water is just over $1 per 1000 gallons which is comparable with that obtained from distillation plants. In Sweden, in 1980, tertiary plants should be universal although at present very few communities possess them. In France, on the other hand, there are many large cities that have no form of sewage works at all. In Britain the debate in the House of Lords in 1970 revealed that as many as three out of every five local authority sewage works were producing sub-standard effluent and the amount of effluent being poured *directly* into the rivers untreated by industry is unknown.

New regulations will probably increase drastically the penalties for polluting the water supplies. At present it is cheaper for many firms to risk being fined rather than improve their standards. In February 1968 slurry from a calcium fluoride-producing plant of Glebe Mines (a subsidiary of Laporte Industries) flooded a part of the village of Stoney Middleton. The result of a prosecution by the Trent River Authority was a £50 fine and 20 guineas costs. Hardly a crippling blow to a major international company, though they are now spending, voluntarily, £250,000 on remedial action: many firms in the same position pay the fine and wait for the next time.

There are, of course, active attempts by companies to try to counter pollution. In the next ten years I.C.I. will spend about £60 million on anti-pollution measures in chemical plants in the U.K. In new plant this can represent about 10 per cent of the total cost. It is the problem of who pays for improvements that is perhaps the most vexed. If industry has to bear the brunt alone, it risks pricing its products out of the world market. Governments, on the other hand, always have something better to do with their money and there have always been more votes in pensions and schools than a clean environment — at least up to now. Pollution-free surroundings are something everyone wants and no-one will pay for.

The River Trent, which runs through the heart of the industrial Midlands, is to be investigated to see if it is possible to clean it. At present it is too filthy for human use. Its main tributary, the Tame, daily pours in 15 million gallons of almost untreated sewage, in addition to the chemicals and other wastes from the industries on its banks. So the Trent starts out filthy before it receives its own effluents. The main problem is that with the increasing population of the Midlands more drinking water is needed, and the river could provide as much as an extra 500 million gallons a day — if it was drinkable. The investigation is backed by the Water Resources Board, the Trent River Authority and a number of other bodies. It is to assess the cost of all possible methods of cleaning the Trent. Every factor will be researched and the final analysis made by computer.

A scheme has already been outlined to clean the Cuyahoga. Over the next five years Cleveland is going to spend $100 million improving present facilities and building a modern sewage treatment plant plus 25 miles of trunk-line sewers. The U.S. Federal Government has outlined a $1.1 thousand million programme to improve the sewage treatment plants in American cities on Lake Erie, and industry has been asked to spend $285 million on waste treatment equipment. There are, however, very few cities that are meeting the necessary requirements.

Canada is also taking action. The Government is committed to a national water quality programme and Federal regulations controlling the phosphate content of detergents came into effect in the summer of 1970. Canada has felt for some time that phosphates from detergents have been a major factor in the pollution of the Great Lakes. The report from the International Joint Commission said 'phosphates from detergents and fertilizers as well as human and animal wastes have been identified as the culprit in ageing of lakes and particularly in fostering the growth of algae.' Since the manufacturers have not responded to gentle persuasion, they are going to be forced to intensify efforts to find a replacement for these phosphates. One already exists. Dr Philip Jones of Toronto University has been experimenting with detergents based on sodium nitrilo-tri-acetate (SNTA) which was originally developed in Sweden. He has found that the SNTA is easily degraded by bacteria to give residues that are neither toxic nor provide nutrients for algae. Manufacturers in the U.S. are already adding SNTA to some of their products.

On the other hand, the American Chemical Society in its report* did not consider that eutrophication could be controlled by limiting the use of phosphate detergents and of phosphates in agriculture. And in Britain the Ministry of Housing and Local Government reported that there had been no increased difficulty from algae growths due to the effects of detergents. The lack of unanimity among scientists, Canadian, American and British, indicates the complexity of the problem of eutrophication. There is not even a set concentration of nitrates and phosphates at which algae blooms will not occur. With a problem as pressing as this, however, with our lakes and rivers at serious risk of destruction, every measure that could retard the ageing process should be tried.

* *Cleaning our Environment. The Chemical Basis for Action*

DEAD LAKES, DYING RIVERS

The increasing use of fertilizers is having adverse effects not only on the water but also on the soil. For too many years fertilizers have been concerned with increasing productivity only and not with replenishing the soil. The result is that soil quality is deteriorating. Soils that have been treated with chemical fertilizers over a long period loose their organic humus content, which is important not only for the soil's fertility but also because it helps prevent erosion. Prior to the modern era in farming, the soil was kept in good condition by the natural processes of plant decomposition aided by manure from the animals that grazed the land. This manure nourished the plant life which in turn fed the animals, and a stable ecological balance between land, plants and animals was maintained. The practice of the traditional Norfolk four-course crop rotation meant that fields were left fallow once every four years and so maintained stable, fertile soils.

The modern methods of intensive farming, particularly in the U.S., have meant that some farms have only livestock; and animal manure, once a valuable by-product of the farm, is now an expensive nuisance. Others, conversely, have none at all. More and more land is being turned over to cereal production or other non-animal orientated systems with an increasing dependence on fertilizers. While those that have been used to increase plant yields have mainly contained nitrogen, phosphate and potash. The other minerals that a plant needs have to be extracted from the soil. The soil thus becomes poorer and poorer in these minerals. Given the need for increased food production, the use of fertilizers in modern farming is obviously essential. But the soil itself must be treated with more respect. It is not indestructible. Unless both the nutrients taken out by the crops and the humus content are replaced, new dust bowls may be created.

The heated effluent being discharged from power stations into rivers and coastal water is another source of danger to the environment. This is known as thermal pollution. Normally game fish, forage fish, plants, insects and dissolved nutrients all live in harmony with each other. Each species is kept in check by the other and there is enough food, air and water for all. Heated water from power stations disrupts the functioning of the whole of the aquatic ecosystem. It kills the normal vegetation and stimulates the growth of algae which deoxygenate the water. Consequently the whole delicate balance is upset. While there is little evidence that fish have been directly killed by thermal pollution, the effects of the heated water on fish metabolism are dangerous in the long run. Fish metabolism speeds up with increasing heat; this increases the fish's need for oxygen and the fish breathes faster. At the same time the heated water from the power station contains less oxygen than the original water drawn from the river. The temperature of the water affects the appetite, digestion and growth of the fish. There are also critical temperatures above and below which fish will not reproduce. In the U.S., the Federal Water Pollution Control Administration consider that fish cannot live in waters hotter than 93°F. Nuclear power stations can raise water temperatures by as much as 16°F. The Federal Government has laid down the limits to which the temperature of the water may be raised: for inland water, approximately 5°F above the normal with a ceiling of 86°F, to 96°F for short periods in certain areas. Trout and salmon streams cannot be raised above 55°F and coastal waters can only be raised by 4°F in the winter and 1½°F in summer. So far thermal pollution has not proved a serious problem but it could if not checked. It would be a pleasant surprise if, for once, measures were taken to avert a potential danger before it has had a disastrous effect.

Nevertheless, the growing need for electricity means that it is essential that more nuclear power stations be built. They provide cheap, smokeless power but need much more water for cooling purposes than the conventional fossil fuel installations. They waste 60 per cent more energy than the conventional plants, all of which is dissipated as heat. Vast quantities of water from the rivers, lakes and seas are needed to cool the heated reactors. It has been estimated that by the turn of the century, in the U.S. alone, nuclear power plants will be producing 1.2 million megawatts of the nation's total electricity output of 1.8 million; using water to cool the condensers will heat the equivalent of a third of the yearly fresh-water runoff in the U.S.

Some method of using all this waste heat must be devised. Experiments are taking place in Oregon in the use of the hot water for farming. The water can be used to heat the soil which in the colder States can protect plants from the frost and stimulate their growth. Another possibility is to construct joint nuclear power/desalinization plants which would use the waste heat from the power stations to help the evaporation in the desalinization process. This kills two birds with one stone; not only producing cheap power but also clean water.

The seriousness of the problem of river and lake pollution has only recently been fully appreciated. This chapter has underlined the fact that it poses a two-fold threat of disturbing urgency. The first is to Man's water supplies for human and industrial consumption. With world population doubling — at the present rate — every 37 years, it is vitally important to ensure pure drinking water for the urban communities where the majority will be concentrated. The second threat is to fish which, because of the rise in population, will be an increasingly vital source of food. The solution to the problem rests largely on how much money is spent on countering it. Industry can return water to rivers after use as clean as it was originally; but it means installing adequate water purification plants which is expensive. Sewage can be cleaned and purified until it is fit to drink, but that too is expensive. Man has the means to ensure a pure water supply and restore life to dead rivers and lakes. Has he the will?

III

THE SEA, SEWER TO THE WORLD

The seas have been treated as a dustbin of infinite capacity for thousands of years. Our ancestors had some justification for this belief, we have none. Their garbage was not only mainly organic in origin, but also limited in quantity and so the oceans were more than capable of breaking down the materials dumped in them. Today the situation is entirely different. The sheer volume of garbage that now finds its way into the seas is a threat in itself. More important than the quantity is the quality of the rubbish — often highly toxic and, being completely unaffected by water, indestructible. There are a host of industrial chemicals like mercury from pulp and paper mills and lead from cars (since tetraethyl lead was first introduced into petrol as an anti-knock agent in 1925, the lead concentration in the Pacific has increased over 10 times and with an element so toxic this is particularly dangerous); there are the pesticides, polychlorinated biphenyls (used in the plastics and other industries), sewage, oil, radioactive waste — an endless list of alien chemicals and materials which are harmful to the marine environment and with which we pollute it in increasingly large amounts.

It may seem far-fetched to believe that the self-purifying processes of the seas could ever be

POLLUTION: THE WORLD CRISIS

overloaded; but when one considers what is happening to the Baltic the possibility no longer seems so remote. The Baltic is a dying sea. Poisoned by DDT, mercury and oil, contaminated by the industrial waste, particularly from the wood-pulping industries, and municipal sewage of Sweden, Finland, Russia and Poland, there are now many areas where all fish life is dead. What fish are caught in the Baltic have been so contaminated with DDT that they have had to be condemned.

The report issued in March 1970 by the International Council for the Exploration of the Sea says that there is a risk that it will 'turn into a lifeless "oceanic desert" like the Black Sea'. Seals in the Baltic contain ten times the DDT concentration of North Sea seals, and the reproductive cycle of fish has been affected. The phosphate concentrations are now three times as high as 15 years ago. Below the Halocline (the depth at which there is a sharp change in salinity, impeding the exchange of water below and above), there are 400,000 tons of phosphates and this is being added to at the rate of 16,000 tons a year. One result of this tremendous increase is that the concentration of poisonous hydrogen sulphide is rising and that of oxygen falling. When the plant life that flourishes on the phosphates dies, it sinks to the bottom and absorbs oxygen as it decomposes. These decomposing plants produce the hydrogen sulphide.

The situation in the Baltic is aggravated by the fact that it is a relatively closed sea. Studies by Svante Oden, a Swedish scientist, have shown that the water takes about 30 years to flow through it. If some action is not taken to prevent further pollution, the whole of the Baltic could die. Sweden is urging Russia, Denmark, Poland, Finland and East and West Germany to co-operate with them in taking measures to combat the pollution; but East-West relations complicate the issue and little has yet been done.

It is vitally important that the marine environment should not be polluted, especially the coastal and estuarine waters. It is in coastal waters that many species of fish breed, protected from the menace of sharks and other marine predators. Shellfish (clams, oysters, scallops, mussels) and crustacea (shrimps, lobsters, crabs) thrive in the estuaries. In the U.S. more than a tenth of the 10.7 million square miles of shellfish-producing waters are now unusable because of pollution. They have been poisoned by a mixture of industrial waste, municipal sewage and toxic chemicals and often their habitat has been destroyed by large engineering operations — harbours, deep-water canals and land reclamation projects. San Francisco Bay once provided 15 million pounds of oysters and 300,000 pounds of clams a year. Neither can be found today. Shrimps are still harvested, but only 10,000 pounds instead of six and a half million annually. In the words of the *Unesco Courier,* of March 1969, San Francisco Bay 'exists largely as a giant cesspool, a garbage dump . . .'.

The shellfish beds in Moriches and Bellport Bays on Long Island, New York, were closed in 1967, because sewage from duck farms in the vicinity had contaminated and killed the shellfish. Newly-built sewage treatment plants have allowed them to recover, and in 1970 about half the beds were re-opened.

Where shellfish have survived they have often been rendered uneatable. An increase in the number of cases of hepatitis in New York City was due to local oysters. The oyster meat was greenish and had an oily taste, because the bay in which they lived was contaminated with oil and copper wastes. They are no longer eaten. A more serious case of poisoning occurred among Japanese eating the oysters of Minamata Bay: in the ten years between 1953 and 1963 some 105 cases of a serious nerve disease were reported ending in either death or severe disability. The cause was eventually traced to the bay's shellfish. They contained unusually high concentrations of mercury which poisons the human nervous system. The cause was a chemical plant that was discharging mercury wastes into the bay. Initially this was in very small quantities but through biological magnification (Chapter V) the shellfish concentrated it in their bodies until dangerous levels were reached. The chemical plant has now stopped dumping its wastes into the bay, but the shellfish are still uneaten.

It is not only the static shellfish that suffer from pollution: even fast swimming fish cannot excape contamination. Tuna caught in the Mediterranean off the oil-refining centre of Fos, in southern France, had to be thrown back because they smelt of petrol. The sea near Marseille, Nice and Fos is seriously polluted by the raw sewage and industrial waste that is dumped into it. Further along the coast an aluminium company at the resort of Cassis pours its waste — the red mud of Cassis — into the sea, and the area is now lifeless.

Uncontrolled use of the sea as a rubbish bin is taking its toll of birds too. Polluted dredged-up muck dumped off Rotterdam, in Holland, was responsible in 1964 and 1965 for widespread deaths among a colony of Sandwich terns hundreds of miles away. The birds were killed by a pesticide that had been washed out of a chemical plant into the mud off Rotterdam's inland waterway. This mud was dredged and dumped offshore where currents carried it out to sea. The birds fed off fish that had been contaminated, and died.

Dredged mud has helped create a 20 square mile dead sea, eight to twelve miles off Long Island, U.S. Sewage sludge and dredged mud has been dumped there for the past 40 years but the noxious effects have grown noticeably worse in the past five. The U.S. Marine Laboratory at Sandy Hook said in a report released in February 1970 that the dead sea was spreading towards the New York and New Jersey beaches. Even sea worms no longer live in this area and the fish that are caught are diseased. They are suffering from fin rot caused by bacteria which are about the only life that is nourished by the pollution. The tails of some fish are disintegrating. Not even plant life survives. The reason is that the dredging spoils contain pollutants, many toxic and others that deoxygenate the water. Under this dual attack nothing can live. The Food and Drug Administration have recommended that all the waters within a six-mile radius of the contaminated area be immediately closed to shellfish harvesting. There is a risk that people who eat the molluscs will get hepatitis.

In this one small area five million cubic yards of sewage sludge — the heavy muck left in the plants after the sewage has been treated — and six million tons of dredging spoils have been dumped each year. In addition raw sewage is poured into the Hudson and East Rivers and washed out to sea. It has been estimated that even if dumping were stopped immediately it would take at least a decade for the currents bringing clear water to cleanse this area. The acting director of the Marine Laboratory, John Clark, has proposed that the dumping grounds should be moved beyond the 20-fathom depth mark, because that area was richest in marine life. There has been the usual battle between cost, convenience and conservation since this would involve hauling the muck 30 miles out to sea. Some New York congressmen have gone even further and called for the site to be moved 100 miles out into the Atlantic. But even this would only be a short-term solution. What are really needed are improved processes for dealing with the sewage and dredged muck which would mean that little or nothing has to be dumped into the sea. Unless Man wishes to destroy a major food source, the nurseries of the seas must be protected.

THE SEA, SEWER TO THE WORLD

At the same time as Man is destroying or decreasing the sea's natural productivity, governments all over the world are experimenting with the 'new science' of mariculture — fish farming — with the intention of adapting the techniques of agriculture to the sea, greatly to increase the yield of the oceans. In Japan more and more coastal waters have been turned into fish breeding farms. A variety of breakwater fences and embankments made of steel pipes and other materials have been introduced to keep new 'sea pastures' safe from the strong seasonal winds. In these pastures fishermen are cultivating bream, bass, conger, prawns, crabs, octopi, oysters and scallops. Oysters have long been cultivated in Japan, and Australia produces some 60 million oysters a year from its farms in the estuaries of New South Wales. In Britain the White Fish Authority of the Ministry of Agriculture has carried out a number of successful experiments in fish farming. Sole and plaice have been hatched and reared to market size in about half the time it takes in the sea. The fish were grown in warm sea water that was discharged from the Hunterston nuclear power station in Ayrshire. They were fed with fish waste. The Authority hopes to improve the stock by selective breeding. Prawns and oysters are also being raised in hatching beds. Eventually all the techniques used in agriculture will be used in mariculture. Warm water from coastal power stations can be used to increase production and the fish themselves can be fed on very cheap foods. This use of hot effluent from power stations is safe, economic and avoids the dangers of uncontrolled discharge into unsuitable environments.

It is this vast expenditure on setting up fish farms in coastal waters contrasted with the simultaneous destruction of natural fish nurseries — deep sea, coastal and river — that is so ludicrous. It is essential to tackle both problems of increasing production and decreasing pollution at the same time to avoid the waste of natural, human and financial resources. For this reason the rivers where so many fish species — salmon, shad, striped bass and sturgeon — breed and the estuaries where plaice, flounder, smelt and many others breed, need careful study and vigorous protection.

Estuaries are very complicated bodies of water. Rivers do not just flow straight out into the open sea. They and the nutrients they carry are trapped by currents, thus providing a very rich source of food for the young fish. Unfortunately these currents also trap the pollutants which become concentrated and poison the estuarine life. A further menace to the equilibrium comes from dams. By decreasing the available amount of fresh water, they upset the essential proportion of fresh to salt water.

We are still ignorant about the mechanics of estuaries. While inter-relationships on land have been studied for many years, the sea has been neglected. The U.S. is now studying Chesapeake Bay, Maryland, in detail. An hydraulic model is being made which reproduces the tributaries (there are about 100 of them) that flow into it, the Atlantic tides and the main sewage flows. Thus they should be able to predict the results of any new sewage works, dredging operations or other changes to the environment. Once the complexity of the life in coastal waters is better understood they can be used both as an industrial resource and as fish breeding grounds without destroying one for the sake of the other.

It is not only the coastal waters that are polluted. There is a distressingly long list of deep sea pollutants. Nuclear tests have meant that all the oceans are contaminated with radioactive fallout. More than a million tons of oil are poured into the seas each year. Since the Second World War the sea has also been used as a general dumping ground for chemical warfare and radioactive wastes. Immense stockpiles of gas were sunk in the Baltic and there are vast radioactive waste dumps off the Atlantic and Pacific coasts of the U.S. and also in the Mediterranean. The containers in the Baltic have now rusted, and in August 1969 fishermen were injured by the mustard gas that had leaked out. The other containers are by no means imperishable: should they leak the results could be disastrous.

Ocean currents carry DDT and other pesticides all over the world; and now a new menace, the polychlorinated biphenyls (PCBs), is being detected throughout the marine environment. During August and September 1969 reports began coming in of large numbers of dead sea-birds being washed up on the coasts of Northern Ireland. This turned into one of the worst disasters of its kind since records began 200 years ago. Throughout September dead birds were reported drifting in to the coast from Ayrshire in Scotland down to north Wales and the Isle of Anglesey. By mid-October over 10,000 birds had been collected. Guillemots were the worst hit but gannetts, razorbills, shags, cormorants and the puffin were all affected. The reasons for these mass deaths are still not fully clear, but PCBs were implicated.

It now seems that these waste products of the plastics, lubrication and cosmetics industries have been polluting the environment for many years, although how and where they get into the biosphere is not known. They have been detected in fish all over the world and were found in high concentrations in the tissues of the dead sea-birds, particularly the guillemots. Although it is not certain that the PCBs were the direct cause of the deaths they were certainly a contributing factor. This disaster may well prove to be a warning of yet another major pollutant of our environment. How many others there are still to be discovered is an awesome question.

Indirect action by Man is decreasing the number of fish nurseries; direct action is killing off the potential parents. Some species, such as the whale, have been so greatly overfished that they are in danger of extinction. Although whale fishing quotas have now been fixed, it will be many years before a stable population is rebuilt. The ban on the fishing of blue and humpback whales in the North Atlantic was renewed for another three years at the meeting of the International Whaling Commission in May 1969. The blue whale has been so overhunted in the past that it is still in danger of extinction. They used to form the major part of any catch, but are now rarely sighted and these speciments are usually underlength. Mature adults can reach 96 feet in length and weigh an estimated 120 tons. Fin, sei and sperm whales continue to be hunted, although the quotas — while still not as low as conservationists would like — are decreasing yearly. The quota in the Antarctic for the 1969-70 season was fixed at 2,700 blue whale units (one blue whale unit *equals* one blue whale *equals* two fin whales *equals* two and a half humpback *equals* six sei) and this is 500 units less than the previous season.

It is possible to fish the seas in a scientific manner — always leaving sufficient behind to ensure that the stock remains abundant and an inexhaustible source of food is safeguarded. There is a point of 'maximum sustainable yield' beyond which a fish species will begin to decline, thus yielding less and less each year. Already stocks of tuna, seals and sea turtles in all oceans are being overfished; herring, cod and ocean perch in the Atlantic and sardine and anchovy in the Pacific are all on the decline.

One species which is rapidly decreasing because it is being attacked through the whole of its life cycle is the salmon. Many of their spawning grounds, far up inland rivers, have been destroyed by pollution and their passage up river blocked by dams. This situation is now being remedied because the salmon is too valuable a commercial and game fish for it to be allowed to disappear. In North America rivers are being cleared and dams fitted with salmon ladders and many other countries are taking similar steps to ensure that the salmon can breed freely. But once the salmon reaches open seas it is now in even greater danger. The recently discovered feeding grounds off the coast of Greenland are being greedily exploited by the Danes. The Danish catch off Greenland has jumped from 127 tons in 1961 to 2,000 tons in 1969. This could have a disastrous effect on the salmon population in other waters. The Danes are ignoring the objections of other sea-going nations and may well endanger the whole salmon population to gain a short term advantage. They argue that there is not enough evidence to show that the decline is due to increased salmon fishing. They claim that pollution and the fish disease UDN (ulcerative dermal necrosis), which is affecting the rivers of England, Scotland and Ireland, are causing the decrease. Whatever the cause, their logic is seriously at fault. Increased deep-sea fishing is no way to preserve what numbers are left. If the international call of April 1969 for a ten-year ban on salmon fishing on the high seas were obeyed, the fish population might have time to recover. Yet the ban has not materialised.

The food resources of the oceans must be protected against both the greed and ignorance of Man. Now that the value of coastal waters is recognised they must be protected against further abuse. Since the sea cannot degrade many of the synthetic chemicals that Man throws into it with such abandon, other less persistent and poisonous compounds must be developed. It is no longer sufficient to make an effective pesticide for use on the land without considering all its side effects once it is released into the biosphere. It is no longer enough just to hope for the best. Both the land and sea have an important role to play in the maintenance of life on this planet. We ignore the needs of either at our peril.

IV

KILLER OIL

If seven maids with seven mops swept it for half a year. Do you suppose, the walrus said, that they could get it clear. I doubt it, said the carpenter, and shed a bitter tear.
　　　　　Lewis Carroll, *Through the Looking Glass*

On March 18 1967 the *Torrey Canyon* piled up onto the Seven Stones Reef off Cornwall and during the next few days 100,000 tons of crude oil poured into the sea. The rest of the 120,000 ton cargo was eventually set alight with high explosives and napalm, but not before a giant oil slick 35 miles long and 15 miles wide threatened to ruin 1000 miles of beaches in Cornwall and Brittany and destroy marine and wildlife on an unprecedented scale.

8,000 oil-covered birds were swept up on the Cornish coast alone and it has been estimated that in all 50,000 sea birds died. The razorbill population was reduced from 2,000 to 1,000 and guillemots from 500 to 100. Whole colonies of nesting birds were destroyed and in Brittany the puffin population, which was reduced from 5,000 to 600, has still not recovered.

One year later the Marine Biology Laboratory, Plymouth, issued a report, *Torrey Canyon — Pollution and Marine Life,* which revealed that it was

KILLER OIL

not the oil that caused the greatest destruction to marine life but the detergents used to clear it. Many parts of the Cornish coast are still a marine desert. Starfish, crabs, lobsters and rockfish were all dead. The detergents also killed the algae which feed the plankton the baby fish eat; so the fish population has retreated to deeper waters. The report estimated that it could take three years for the normal population to be restored.

The French sank the oil using a chalk-stearate mixture and they have been much less severely affected. Once the oil sank to the bottom it was broken down by the natural action of bacteria and caused very little harm to fish populations. The oil was cleared from Brittany beaches by bulldozer and shovel which again minimized the destruction. Both the British and French Governments spent large sums of money clearing the oil — the British about £5 million. Getting compensation from the ship's owners proved a difficult task. A sister ship of the *Torrey Canyon* was seized in Singapore and only released after a large bond was deposited. The claims were eventually settled out of Court in November 1969, when both Governments received £1.5 million each from the owners, Barracuda Tanker Corporation of Monrovia — a U.S.-controlled firm — and the charterers, the Union Oil Co., of California. A lawsuit could have dragged on for years with no guarantee that even then all their claims would be met, since the international legal system over liability for oil pollution is still not clear.

POLLUTION: THE WORLD CRISIS

In January 1969 one of the offshore oil rigs at Santa Barbara, California, operated by Union Oil, blew out. Every day 21,000 gallons of crude oil boiled into the sea through faults in the sea bed. The initial attempts to stop the blow-out were unsuccessful and it was 12 days before the well was plugged with huge amounts of a chemical 'mud' mixed with cement. Over 250,000 gallons had escaped. Vast amounts of crude oil hovered offshore for five days while the winds kept it out at sea; then the wind changed and the beaches of the resort of Santa Barbara were fouled by the sticky, black gunge. The slick eventually spread over 200 miles and affected the coast as far south as the Mexican border.

The lessons of the *Torrey Canyon* disaster had been learnt and, except for the water-soluble dispersant used to treat the oil slicks, chemicals were not used. Straw, talc and pearlite were spread along the beaches to soak up the oil, the straw was then raked up and bulldozed and carried away in trucks. Booms were placed around the drilling platform to try and contain the oil, but due to poor weather the efforts were unsuccessful. The full ecological consequences of the catastrophe have not yet been evaluated but California's Department of Fish and Game estimates that 50 per cent of Santa Barbara's bird population has been destroyed. A survey along a 30-mile stretch of beach revealed 200 grebes compared with a normal population of 4,000 to 7,000. The effects on marine mammals do not seem as great as initially feared. But grey whales migrating

KILLER OIL

from the Bering Sea to Baja, California, now avoid the Santa Barbara Channel. Some damage was reported to the local herds of seal and sea lions. The sharpest threat occurs immediately after birth when the pups are three days old, the cows leave them on the beaches and swim out to sea to recover their strength and gather food. By the time they return to the beach, the pups are often so completely covered in oil that they are unrecognisable and abandoned to die.

The U.S. Geological Survey issued a report in 1970 explaining what took place. The Union Oil Company had drilled the well 3,500 feet below the ocean floor through a number of oil reservoirs lying beneath a thin layer of unusually porous rock; the deeper levels contained oil at a greater pressure and it was this that rushed to the surface of the topmost reservoir and penetrated the porous rock through various faults in the structure.* It now seems that the only way to relieve the pressure in the basin is to continue drilling until all the oil is removed. This could take two years during which the likelihood of other blow-outs is high. This would effectively mean the end of Santa Barbara as a tourist area. Swimming in oily waters and sunbathing on sticky beaches hold few attractions.

* *Oil blow-outs are not at all uncommon even under the most favourable conditions. The Department of the Interior estimate 2.5 per thousand wells suffer from this fault and drilling in unstable areas can be expected to increase the frequency.*

The cleaning-up operations were a very expensive business. The oil companies were made responsible for cleaning up any pollution caused, by order of the Secretary of the Interior in February 1969. Quite how much these operations cost is difficult to assess but it could be as much as $5 million. In addition there are some $2,000 million worth of lawsuits pending against the companies and the Federal Government. It took nearly a year before the beaches could be declared reasonably clean and it has been estimated that the local tourist trade lost over $11 million during 1969.

But for Man's insatiable appetite for oil the Santa Barbara disaster need never have happened. The oilfield off the Santa Barbara coast was well-known as being in a geologically unstable area. There has been a long history of natural leakages of oil but despite these warning signals the Federal Government granted the oil companies leases to drill. This was worth $1.6 thousand million to them in revenue in 1968 and the lease that included the ill-starred Platform A brought in $61.4 million.

The residents of Santa Barbara formed a society called GOO (Get Oil Out) after the blow-out, aimed at stopping all further drillings in the area. But their chances of success are slim with the financial power ranged against them. The original leases brought $603 million into the Federal coffers and revoking them would be a very expensive and complicated matter. Although drilling operations were initially shut down they were resumed after the presidential scientific panel formed to investigate the disaster recommended that drilling be continued to relieve the pressures in the oil pool. Since then nearly 60 new oil wells have been drilled in the Channel and the oil companies, Union, Gulf, Mobil and Texaco, are producing approximately 30,000 barrels of oil a day – $90,000 worth.

It seems unlikely that with so much money tied up in the oil rigs and in Federal leases, GOO and their supporters will be successful in getting rid of the oil companies. Senator Edmund Muskie, a leader in environmental legislation, is urging the Federal Government to revoke the leases, compensate the lessees and make the Channel a national oil reserve, to be tapped only in an emergency. A happy dream. The oil companies are already complaining that a Federal Bill before a Senate committee, which would penalize offenders up to $125 a ton up to a maximum of $14 million with unlimited liability in case of negligence, is too harsh. They can be expected to fight it. Since the oil lobby in the U.S. is extremely influential, it is more than likely that much desirable legislation in this vital field will be blocked.

The repercussions of the Santa Barbara affair have been felt 7,500 miles away in Australia. Plans to drill for oil near the Great Barrier Reef have been postponed until a government survey of the probable effects on the environment has been completed. The consequences of the drilling on the Great Barrier Reef alarmed conservationists who threatened to organize a union boycott of a floating rig if it was brought in to do some test drilling off Queensland.

The Great Barrier Reef is one of the natural wonders of the world and, rich in colourful marine life, is one of Australia's greatest tourist attractions. It is also an area of unique ecological value and great scientific interest that has been described as one of the world's great biological treasure houses. Stretching for 1,250 miles along the Queensland coast it is a valuable source of marine food and also helps protect the coast. An accident of the proportions of Santa Barbara would have disastrous consequences. Oil would coat the reef and kill the living coral. Once the coral dies, the sea can slowly erode away the reef. Not only would Australia's east coast be vulnerable but this habitat for thousands of species would be destroyed.

There has been a long and bitter argument between the Federal Government of Australia and the State of Queensland as to who has jurisdiction over the reef and the waters between it and the Queensland coast. While the High Court has ruled that the Federal Government has jurisdiction over sea areas beyond the low water mark, action by Mr Gorton, the Australian Prime Minister, had been blocked by the State Government trying to maintain its autonomy. Queensland stands to gain vast sums of money if oil is found in commercial quantities and is understandably eager to welcome the oil companies. However the Federal and State Governments have reached agreement on the need for a scientific survey and the Great Barrier Reef has won a temporary reprieve.

KILLER OIL

Another unique ecological area that is threatened by oil is the northern slopes of Alaska. Oil has been found there in vast quantities and the State of Alaska has been enriched to the tune of $900 million in 1969 alone. Oil companies are eagerly exploiting this new reserve. Since the Arctic coastline is frozen for much of the year, a trans-Alaskan pipeline carrying the oil from Prudhoe Bay to the Port of Vallez on the Pacific Ocean is planned. This pipe, four feet wide and nearly 800 miles long, would carry the heated oil across the frozen land. A heated pipeline buried in the soil would melt the permafrost and could lead to serious erosion. At least 40 per cent of the pipeline may have to be carried above the land on stilts high enough to allow the animals of the Arctic—particularly the caribou—free passage. For the proposed line of the pipe crosses two major migration routes of the caribou—the staple food of the Nunamiut Eskimos.

Because of the intense cold the consequences of any accidental oil spillage in the Arctic environment are infinitely more serious than in warmer climates. In temperate zones, the volatile components of the oil are fairly rapidly vapourized and the oil is degraded or broken down by natural processes quite rapidly. In the sub-zero temperatures of the north the processes of biodegradation are much slower and oil can remain unchanged, defiling the land for years. The huge girth of the pipe and the high speed of oil flow means that in the event of a fracture vast areas would quickly be covered with hot oil (it will be maintained at 80°C).

The normal construction activities that are part of any oil drilling camp also destroy the land. Garbage and sewage are normally broken down by the action of bacteria; these processes are so slow in the Arctic that the oil companies have been bulldozing shallow lagoons into which the wastes are dumped. This is neat and simple but could lead to general contamination of the water-table as the wastes seep through the spongy tundra. To minimize ecological damage some of them have used helicopters to transport their equipment to avoid tearing the land to pieces.

After the experience at Santa Barbara, the Federal Government is showing a keen interest in the possible environmental consequences in Alaska. The Secretary of the Interior, Walter Hickel, has set up a special Arctic Task Force to devise guidelines to control development and minimize the damage to the land. The pipeline must have automatic pollution-detecting devices and many shut-off valves.

The alternative route for Alaskan oil is via the North-West Passage. If one of the giant oil tankers broke up there, the catastrophe would be appalling. On her trial voyage the 114,000 ton tanker, *Manhattan,* sailed through the North-West Passage to test the feasibility of this route. Two huge holes were found in the hull 40 feet below the waterline. Even in summer the *Manhattan* needed the assistance of icebreakers to get through, and it may prove that the scheme of carrying oil through the Passage is uneconomic. It is a sad reflection that only economic interests could put paid to an idea which common sense should have dismissed long since.

The species that suffer most severely and obviously from oil pollution are the sea-birds. It is now quite commonplace to open one's newspaper and be greeted by the sight of yet another pathetic, bedraggled, oil-soaked bird — the inevitable victim of any oil slick. Millions of them are lost each year, many belonging to endangered species. In South Africa the blackfooted penguin has been severely reduced as a result of oil pollution. Other birds that are threatened include guillemots, razorbills and puffins which are especially vulnerable because of their low reproduction rate.

Research is currently going on into the most effective methods of saving oil-soaked birds. The Newark University zoology department in Britain is to carry out a five-year research programme into finding a cheap, effective way of treating contaminated birds. Many die while in the hands of those trying to help them. In California, after the Santa Barbara oil disaster, it was found that often birds died because they were deficient in thiamine. The reason for this was that in capitivity they were only being fed on a diet of raw fish which did not supply this essential vitamin. The cure is quite simple: massive doses of vitamin B1 in the diet. The birds are often starving as well as badly shocked when they are captured, all of which militates against their recovery even when they have been cleaned. An oil remover has been developed in Sweden that is claimed to be able to remove the contaminating oil without removing the birds' natural protective oil. This is essential: birds washed with ordinary detergents, when returned to the wild, promptly become waterlogged and sink. Even with the most effective of methods of rehabilitation thousands of birds are bound to die while oil continues to pollute our environment.

Stopping the pollution at its source, at the rigs that drill for it and the tankers that carry it, is clearly the best safeguard. With the development of the 200,000 ton supertankers, the dangers increase. Most of the oil slicks that wash up on the shores of Britain are caused either by accidental spillage or by ships washing out their oil tanks at sea. Either way it is very difficult to identify the culprit. After a tanker has off-loaded its cargo a considerable amount of heavy, sticky sludge remains in the tanks. The major oil companies now wash this oil out in harbour and pump the washings into refineries. However there are still a number of tankers that adopt the old method of washing-out at sea. The regulations in force in 1969 meant that it was illegal for any ship to jettison oil within territorial waters and also within the English Channel, the North Sea, the Western Approaches, the Baltic and the Irish Sea. Thus, in theory, there should have been no oil pollution in the whole of northern Europe. This was manifestly not the case. In Britain alone at least four oil slicks are sighted each month; no sooner has one been cleared away than another looms up.

KILLER OIL

In October 1969 the Inter-governmental Maritime Consultative Organization (IMCO) approved more stringent controls against oil pollution by ships. Ships are to be prohibited from discharging tank washings with more than 100 parts of oil in 1 million parts of water in any sea area. No residues are to be discharged within 50 miles of any shore. But ratification of these amendments to the international conventions could take three years, since each government has to approve them; meanwhile the damage to our environment continues unabated.

The problem of compensation for damage by oil pollution has also been dealt with. In November 1969 the International Conference on Marine Pollution agreed on a convention dealing with civil liability for oil pollution damage. Ship owners will be held solely responsible for oil pollution arising from accidents at sea regardless of whose fault the accident was, except in the case of natural catastrophes or government negligence of port maintenance. The oil companies have already set up a voluntary oil pollution plan — Tovalop — under which liability would be accepted up to $10 million per ship for any catastrophe. Over 50 per cent of the world's tanker fleet have agreed to this.

The difficulties in identifying culprits who wilfully or accidentally break these regulations remain. In any one year more than 25 million tons of crude oil are shipped into the three refineries in the Thames Estuary alone, and the majority of ships that visit the Port of London carry oil in their bunkers as fuel.

Thus there is no shortage of suspects when an oil slick is sighted. In the English Channel all ships' masters, Trinity House pilots, coastguards and airline pilots are ordered to report any slicks. It is thus unlikely that a slick could remain undetected for long. (This is not the case in less crowded shipping areas where often the oil slick is not sighted before it reaches the beaches.) The authorities have to know when and where the oil was spilled, what ships were in the area at that time and what type of oil was concerned. There are now a number of sophisticated detection techniques to answer the last question. These were applied during the search for the ship responsible for fouling the beaches of Southend during the Whitsun holiday in 1969.

Samples of the oil from Southend were analysed in a gas-liquid chromatography machine. Oil is made up of number of different components (fractions) — crude oil from different areas of the world has different compositions and crude differs from bunker oil. When placed in the machine the oil is vapourized and the different fractions boil at different temperatures, thus each type of oil produces a characteristic chromatogram. Unfortunately identification of oil in oil slicks is not that simple because the oil is changed by exposure to sea, sun and wind. However, the oil companies have made many different chromatograms of different types of oil under different conditions with which any new chromatogram can be compared for similarities.

The difficulties of a definitive classification of a

given oil sample are clearly illustrated by the Southend case. The Shell laboratories at Thornton identified it as Kuwait crude; the BP laboratories, on the other hand, stated categorically that it was not but was rather some kind of bunker oil. The Port of London Authority summed up that the composition resembled crude oil but did not correspond to any that was being imported into Britain.

Given the time of the first sighting, the prevailing wind and tide conditions, the Port of London Authority knew that the oil must have been discharged on the night of May 24/25 in the Oaze Deep — the main deepwater sealane into the Thames. Detailed records are kept of the ships that enter and leave the Thames and from these it was possible to determine which ships were in the area at the time. Despite all this, the culprit was never identified.

Advance warning of approaching oil slicks and prompt action to disperse the oil at sea could do much to minimise the effects. In 1969 West Germany, Belgium, Britain, France, Norway and Sweden all agreed to warn each other when oil slicks were sighted in the North Sea. This is a useful first step; enforcement of IMCO's regulation should help prevent the process in the first place. However, procedures for dealing with the oil once sighted are still confused. If the oil threatens a busy tourist area the reaction is usually prompt and efficient, but when less inhabited areas are threatened it is another story. In Britain the Board of Trade is now responsible for dealing with the oil at sea but local authorities are responsible for dealing with pollution in their own areas. Warning of slicks is often given by local coastguards and the authorities can ask the Air Force and Navy for help in tracking them. The RSPCA and RSPCB deal with the oiled sea-birds once the slick has struck, although they are not part of the official apparatus.

In practice, advance warning procedures are often inefficient and co-ordination between the official bodies concerned not all that could be desired. Since oil slicks are no respectors of county boundaries, co-operation between the various local bodies is essential. The oil from the tanker *Hamilton Trader* that collided with a ship near the Mersey in April 1969, threatened the north Wales coast, north-west England and south-west Scotland. It was eventually washed ashore in Cumberland and killed nearly 3,000 birds. During much of its journey it was entirely untracked.

While oil pollution continues to be such a threat, it should not be dealt with piecemeal: a co-ordinating body is called for. It would be cheaper and safer to deal with the slicks offshore rather than wait for them to land and this is something that central government rather than local authorities is better equipped to do. Every year over a million tons of crude oil (out of a total world output of around 2,000 million tons) are spilt into the sea — most of it accidentally. The new generation of 200,000 ton supertankers carry up to 60 million gallons of crude oil. With the frequency and extent of disasters rising all the time, national as well as international procedures should be evolved to deal with the consequences.

What is the best method of dealing with an oil slick once it has formed? The British Government seem to favour chemical dispersal of the oil, the French sinking the oil with chalk-stearate or treated pulverized fuel ash so that the natural processes of biodegradation slowly break it down. The French system cannot be used over shellfish areas or fish breeding grounds. The Russians use mechanical methods to mop it up. Detergents are no longer used. The disastrous results obtained by the British after the *Torrey Canyon* affair made it quite clear that these were dangerous. In Britain the Ministry of Agriculture and the Nature Conservancy have prohibited the sinking of oil or use of oil-spill removers in certain areas to avoid damaging valuable marine breeding grounds.

The oil companies are all producing chemical dispersants that are non-toxic to marine life. Corexit is one; it is effective against fresh oil at sea but does not work so well on weathered oil and not at all on the beaches. Dispersol is another; it emulsifies the oil and sea micro-organisms dispose of the emulsion. It has a low toxicity to marine life and birds that are contaminated by the treated oil can be cleaned far more easily than those contaminated by raw oil.

An interesting method was used in Israel in June 1969. A pipeline was sabotaged and oil poured into

POLLUTION: THE WORLD CRISIS

Lake Tiberias. Carpets made of polyurethane flakes were used to soak up the oil and then removed from the water. It has also been suggested that peat could be used in a similar manner, since this has good oil-absorbant properties.

Mechanical methods of dealing with the oil are favoured in Russia. After the *Torrey Canyon* disaster the Russians held a competition for new ideas to combat oil pollution. One of the better ideas is being developed. Rotating shields at the front of a boat scoop up the oil and debris from the water. The oil flows into a settling bath where the debris is removed. The water is pumped into other tanks for further separation and then pumped ashore or discharged, clean, back into the sea. The usefulness of this method is limited by the weather and by the ship's capacity. A ship was used to try and clear some of the slick at Santa Barbara, but however efficient the oil separating process, there is a limit to the amount of water that can be treated.

The Russians also have huge land-based purification plants along the Volga where oil is being drilled. The contaminated water is sucked into the plants where the oil is separated and pure water put back into the river.

Oil booms have been used with some success both to contain the oil in one spot, near the leak, and to stop free-floating slicks from reaching harbours. These, again, are limited by the weather: in rough conditions they tend to break and the oil escapes.

Since oil will continue to be needed for many

generations to come and will have to be transported by sea, we must resign ourselves to the fact that oil is bound to go on contaminating our environment. But there is no reason why it should continue to wreak such havoc with the marine population or continue to foul our beaches. Prevention is better than cure, but when prevention fails, the methods of cure are at hand. Swift action by the authorities whenever an oil slick is sighted could do much to minimize its harmful effects. Enough experience has been gained since the *Torrey Canyon* débâcle for there to be no excuse for incompetent handling of any further disasters. Methods now exist for dealing with oil pollution; it is surely not beyond the wit of man to apply the correct remedy promptly in each case.

V

PESTICIDES, THE INDISCRIMINATE KILLERS

Of all the ways in which Man is upsetting the delicate balance of nature, probably the most widespread and most flagrant is his use of pesticides. To fight plant parasites, human disease and to increase food production, he has assembled a vast armoury of chemicals developed to kill harmful insects. These chemicals have been startlingly successful. Food production has risen enormously. Diseases dangerous to Man, such as malaria, yellow fever, plague and typhus, have been eradicated over a vast area. Unfortunately they have been a great deal too successful, especially the early organochlorines like DDT and dieldrin. Through profound ignorance of the natural environment, they now contaminate the whole biosphere.

DDT was first used during the Second World War. Today it is present in the water, soil and air, stored in human fat, imbibed in human milk and has even been detected in Antarctic snow — thousands of miles away from the nearest spraying operations. Over 1,000 million pounds of DDT has been dumped into our environment since it was first introduced, and we are adding to that store at an estimated rate of over 100 million pounds a year. Its effect on fish and bird life has been disastrous. In the Baltic many fish are too toxic with DDT to be eaten and many bird species such as the peregrin falcon and sparrowhawk are in danger of extinction, because of its effect on their reproductive cycle. As far as Man himself is concerned, perhaps the most disturbing fact is that breast-fed babies are receiving heavy doses of DDT. In North America human milk on average contains over twice the amount allowed in milk for commercial sale (0.05ppm) — in some States as much as six times the maximum permitted dosage. In Australia human milk often contains 30 times as much.

The long-term effects of this alien chemical on our bodies are not yet apparent, yet it does seem likely that introducing large amounts of synthetic chemicals into the body will have harmful effects in the long run. An American report stated in 1969 that some patients who have died of leukaemia, cancer and high blood pressure, had high levels of DDT in their tissues.

One of the most disturbing aspects of DDT pollution is its effects on some plankton — microscopic aquatic plants and animals. In concentrations of as little as 0.01 parts per million (ppm) it can inhibit their photosynthesis by as much as 80 per cent. Considering that 30 per cent of all oxygen in the atmosphere is the product of plankton photosynthesis, the results of this poisoning could be unfortunate. If two tankers of the *Torrey Canyon's* size loaded with DDT were to collide and sink, the contents of their holds could kill many plankton in the Atlantic.

The consequences of these persistent (i.e.

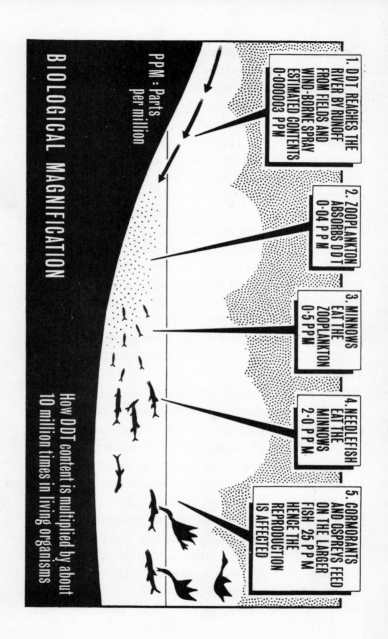

PESTICIDES, THE INDISCRIMINATE KILLERS

long-lasting) pesticides on the whole marine environment have yet to be seen. They are, however, probably already adversely affecting the productivity of the oceans and the oceans are a major food source for the future.

DDT's effect is devastating because of a phenomenon called biological magnification. This is the process by which the concentration of nutrients (and, unfortunately, other chemicals) is increased up the food chain, as one animal eats another. DDT is washed from the fields and forests into rivers and seas. The initial concentrations in the water are very low (0.000003 ppm), but DDT is almost insoluble in water and hence is picked up by the nearest biological material. Plankton concentrate the DDT (0.04ppm) and are then eaten by plant-eating fish (2.0ppm), which are in turn eaten by birds (25ppm and above). Quantities of 100 to 1,000ppm of DDT can kill birds and mammals, and lesser concentrations are sufficient to disrupt the reproductive cycle.

DDT's harmful effects on birds are brought about by a complex process which is still not completely understood. But briefly, it acts on the liver which produces enzymes that effect the sex hormones — testosterone, progesterone and oestrogen. This has a number of undesirable results, chief of which is the effect on oestrogen levels which may be either raised or lowered. Oestrogen controls the production of calcium and this leads to birds laying much thicker or thinner eggshells than normal. Either the chick cannot get out or the egg doesn't survive the hatching period. Both ways bird deaths go up. These observations have been confirmed many times in laboratory experiments.

In America an eagle egg was recently found that had no shell — just a membrane: and in California in 1969 the brown pelican did not breed at all because the eggs collapsed when the birds tried to incubate them. In Britain the peregrin falcon, sparrowhawk and golden eagle have all bred less successfully since DDT and other organochlorine pesticides were introduced. Today they are all seriously threatened species. In America a number of bird species round the Great Lakes including America's symbol, the bald-headed eagle, are in danger of extinction for this same reason. In Sweden the white-tailed eagle is on the decline. Only two of ten pairs nesting in the archipelago of Stockholm have succeeded in hatching during the past few years. All over the world the same story is being repeated.

One of the difficulties in evaluating the dangers of DDT is that other factors often mask its direct effects. In the case of the golden eagle in the Highlands of Scotland, although there was a drop in hatching rates after the introducing of DDT, there seemed to be no drop in population. Further research, however, revealed that because it is a long-lived species, about 20—25 years, it would be at least 15 years before any significant fall-off in numbers would appear, when — incidentally — it might be too late to do anything about it. One sign that these deaths were directly attributable to organochlorines was that in

Scotland, since dieldrin sheepdips were banned in 1964, the rate of successful hatching of the golden eagle eggs, having dropped from 72 to 29 per cent, went back to normal.

DDT itself does not remain unchanged in animal tissues — studies have shown that earthworms turn it into DDE, while slugs turn it into TDE (both metabolites of DDT). Controlled studies at the Patuxant Wildlife Research Centre in Maryland, published in 1969, showed that DDE damaged reproduction in mallard ducks more severely than DDT although, in general, DDE is less poisonous than DDT. DDE has been found widely distributed throughout the environment even though it is not produced commercially. Thus it is not only the DDT used all over the world that damages the environment but also its metabolites.

Of this large range of organochlorines, some are toxic and very persistent like dieldrin and endrin — environmentally the worst; others, like endosulphan are extremely toxic but less persistent. Overall, DDT has probably been the most harmful pollution threat because, though not very toxic, it is persistent and is used in enormous tonnages.

Great numbers of small invertebrates live in the top few inches of the soil. An acre may contain as much as half a ton of earthworms. They play a key part in maintaining the soil's structure and fertility. The Rothampstead Experimental Station in Hertfordshire, England, has been carrying out an extensive research programme into the effects of pesticides on this population which is composed of many different types of animal. The most important are earthworms, woodlice, millipedes, oribatid mites and various insects such as springtails, termites and beetle and fly larva.

They eat the plant residues and work them into the soil, thus increasing its fertility by raising its humus content. This helps aerate the soil and also controls its water-holding capacity. The earthworm is probably the most effective of the natural cultivators: it is also one of the largest and is, therefore, in theory more affected by pesticides than the smaller invertebrates. At first sight, however, the effects of pesticides are not very dangerous; for although the organochlorines, chlordane and heptachlor, do kill them (as do the organophosphorous compounds, though these, being less persistent, are not very damaging), the number of earthworms is not markedly reduced. The real danger is that the earthworms concentrate the pesticides in their tissues and may contain ten times the concentration of the surrounding soil. This can have unfortunate results on animals higher up the food chain. In 1954 Michigan State Univeristy began DDT spraying of elms to kill elm bark beetles which were spreading the Dutch elm disease. It killed the beetles, temporarily, but also succeeded in annihilating the robin. The robin had eaten the earthworms which had eaten the leaves covered with DDT, and the robins died of nerve poisoning. This is a classic example of biological magnification.

PESTICIDES, THE INDISCRIMINATE KILLERS

On cultivated land the direct effects of spraying on the ecology of soil animals and soil fertility are not too marked. What it loses on one level of life, it gains on another. After spraying, though the population of larger invertebrates goes down, this is usually compensated for by a rise in that of mites and springtails which have an equivalent effect on the soil. In uncultivated land and forests, however, the position is more serious, particularly with indiscriminate aerial spraying. Not only does the soil fertility and structure break down because of a direct effect on the soil population itself, but it also poses a serious threat to wild life.

This is well illustrated by a 1969 report from the University of Maine, which shows how long pesticide residues can last in small mammals. DDT was used to control the spruce budworms in the forests of northern Maine. It was sprayed in 1958, 1960, 1961, 1963, 1964 and in 1967 in various areas, at the rate of one pound per acre. Some received more than one treatment but there were large areas that had only one. Mice, voles and shrews were collected and analysed over the nine-year period. It was found that even nine years after only one application there was still some residual contamination in all species. While the level in mice and voles was approaching the pretreatment level, shrews, which are carnivorous, had ten to forty times more DDT than mice and voles and probably will not reach pre-treatment levels until 15 years later. As was to be expected, animals from areas that had been sprayed more than once had much higher levels of DDT. It was also found that various metabolites of DDT were present. Of the total amount of pesticides found in the tissues, mice had 40 per cent DDE and 16 per cent DDD, voles 18 per cent DDE and 28 per cent DDD, and shrews 57 per cent DDE and 17 per cent DDD.

The DDT level found in shrews in the treatment year was high enough to be toxic and certainly high enough to endanger the higher carnivores that eat them. It has been calculated by Dr Norman Moore of Monks Wood Research Station, Hertfordshire, that because of biological magnification a kestrel can die after eating seven mice that have been feeding on grain treated with DDT. Lower levels suffice to affect reproduction adversely. The Bermuda petrel is decreasing in number. They have probably been feeding — via the normal food chain — on contaminated oceanic plankton and have been failing to reproduce. Dead chicks and unhatched eggs have been analysed that have contained organochlorine residues of about 6.55ppm. DDT and the other organochlorine pesticides have also been directly responsible for large numbers of bird, mammal and fish deaths, as was definitively catalogued by Rachel Carson in *Silent Spring*. It is, however, the *indirect* effects of DDT that, it is being increasingly realised, present the greatest threat to the environment.

DDT is a poison and was developed to eliminate undesirable insects and other pests. Unfortunately it destroys beneficial insects as well as harmful. It acts on the nervous system, causing spontaneous and continuous nerve impulses all over the body which lead to convulsions and death. If DDT was excreted at the same time and the same rate as it is taken into the body, there would be little risk. The trouble is that it is absorbed by the body fat and stored. (Luckily Man, unlike birds, excretes much of the DDT he eats.) Slowly DDT in the fat builds up to a level where, when fat is metabolised to meet energy needs at a time of stress, the nervous system is suddenly bombarded with a catastrophically high concentration and the animal dies of poisoning. It is this property of being retained and stored in increasing amounts that represents the chief danger of DDT and the other organochlorines.

The animals at the top of the food chain are not the only ones adversely affected by DDT and similar chemicals: some 700,000 coho salmon from Lake Michigan were condemned by the Food and Drug Administration (FDA) in the U.S. in 1969 because they had unacceptably high concentration of DDT. Michigan then banned DDT spraying in that State.

The tale of the coho salmon is a sad one. In 1833 the Americans finished the Welland Canal joining Lake Ontario to Lake Erie and by-passing Niagara Falls. For the first time lampreys were able to swim up from Lake Ontario into Lake Michigan and Lake Huron. By the 1930s they were wreaking havoc on the fish and, aided by over-enthusiastic fishermen, virtually eliminated the lake's large commercial and sport fish. At the same time alewives greatly multiplied, as the salmon, their natural enemy, was progressively eliminated. But the alewives increased too abundantly, eating out their sources of food: millions were washed up dead on the beaches. In the 1960s it was decided to control the lampreys by pouring a pesticide called TFM into their spawning streams. This worked, and the lakes were restocked with coho salmon, chinook salmon and lake trout which feed on the alewives. By 1968 thousands of fishermen were again descending on the Great Lakes to enjoy their favourite sport. Over the years, however, vast areas of Michigan and other States around the Great Lakes had been liberally sprayed with DDT and other pesticides which had been washed into the Lakes. There, due to biological magnification, it was concentrated in the food chain. The coho salmon was eating the alewives which were eating the micro-organisms living in the mud contaminated by DDT. Result: the coho salmon could not be eaten and 11 per cent of the salmon fry in the Michigan hatcheries in early 1968 died.

Adult fish can live with quite high levels of DDT in their tissues, but the fish embryo dies as soon as it begins to absorb it. In May 1969 the FDA, which fixes the permissible level of DDT in foodstuffs, established five ppm as a temporary level for fish. This meant that no fish containing more than five ppm could be transported across State lines, thus disqualifying 80 per cent of the fish caught in Lake Michigan. Yet again some ecological research done in the initial stages could have saved vast sums of money and needless destruction of life.

PESTICIDES, THE INDISCRIMINATE KILLERS

Another classic case of the misuse of pesticides began in 1949 at Clear Lake, California. In that year, in order to control the gnats, the lake was treated with DDD which is not as harmful to fish as DDT. The concentration was one part in 70 million which was considered adequate to kill the gnats but leave the fish and birds unharmed. Initially the treatment worked, but by 1954 it had to be treated again, this time in a concentration of one part in 50 million, and again in 1957 in the same proportion. By 1960 a large number of western grebes were dying. The cause was found to be pesticide poisoning. They contained up to 1,600ppm of DDD. This enormous concentration was once again the result of biological magnification. Plankton in the lake contained 5ppm — fish which eat plankton contained 40—300ppm — fish-eating fish still more, until at the top of the chain the birds were dying with 1,600 ppm: these situations are typical of what has been happening in lakes and rivers the world over.

One of the most potentially dangerous effects of DDT is on the estuary breeding grounds of marine animals. These are the major sources of the world's fish. This situation is particularly ominous since it is predicted that as the world's population soars (it will double in the next thirty-seven years) Man will have to rely increasingly on the sea for his food.

DDT is carried down the rivers to the estuaries where the currents keep the water circulating. This builds up nutrients for the young fish but also stops

POLLUTION: THE WORLD CRISIS

the poisons being swept out to sea. Its effects here are threefold: first, the baby fish and shrimps are highly susceptible to direct poisoing; secondly, its concentrations in the tissues of organisms on which the breeding fish feed cause a fall in successful reproduction; and, thirdly, the adult fish become dangerous to eat. Thus the whole of the marine environment is at risk. The Baltic is a case in point. It is so situated that it receives the chemicals washed from the land of seven intensively cultivated countries; hence the animals trying to live in and on it are all badly contaminated.

It was the discovery that herrings caught in the Baltic Sea contained high amounts of various chemicals that stimulated Sweden to order a two-year ban on DDT in March 1969. Analyses by four Swedish scientists, published in October 1969, of a wide range of marine organisms from the Baltic showed once again that it is heavily polluted. Both DDT and PCBs were found. That the effects are worse in restricted areas was shown by the fact that mussels from the Baltic and the archipelago of Stockholm had higher levels of DDT and PCBs than those from the west coast of Sweden, more open to the North Sea. DDT concentrations in plaice and cod from the Baltic were higher than in those from the west coast. Even seals from the Baltic contained ten times more DDT and PCBs than specimens from Great Britain, Canada and the Netherlands. The analyses also showed that each link of the food chain, each prey-to-predator step, increased the DDT concentration by at least ten times, and the increase from fish to osprey and heron was 100 times. In the birds the residues were nearly all DDE, while in fish and seals they were about 50 per cent DDT.

It must be acknowledged that since it was first introduced DDT has been responsible for saving millions of lives. It was first used in malaria control programmes in the South Pacific by the U.S. army. Since then it has been effective in controlling disease-carrying insects the world over and has probably saved as many as 10 million lives. Initially its persistence in the environment was an advantage: spraying operations did not have to be repeated and costs were thus kept down. Moreover, the compound was and is used to control agricultural pests: here too, initially, it was highly successful and helped to raise crop yields. One of the main attractions of DDT is its cheapness. But with some 500,000 tons of it circulating in the environment, it has changed from a boon into a menace.

Insects have now evolved resistant strains — about one hundred strains of resistant insects exist, including those that carry malaria, yellow fever and plague. Insects breed very fast — aphids, for example, can produce as many as one generation every fortnight — which means that DDT-resistant strains of 'super-bugs' can be produced within a few months, long before the pests' natural predators can make a recovery. Hence replacements for the organochlorine pesticides are essential if we are to maintain and eventually win the battle against loss of agricultural production and human life.

The criteria for these replacements are that they should be absolutely specific to the undesirable species and non-toxic to other forms of life. There are several fields where this research and development is being carried out. One of the most promising is that of so-called biological control. This uses the principle of 'set a thief to catch a thief'. At the moment there are two main methods: one is to let loose sterile males to upset breeding patterns, the other to introduce the pests' natural enemy in large numbers. The traditional method of controlling fruit-fly in the grape fields of California was to send the children out into the neighbouring lucern fields to collect ladybirds. They put them on the vines and left them to keep down the fruit-fly population. Also on Californian grapes, a wasp has been used to control the leaf hopper, *dikrella*. *Dikrella* is now immune to DDT but the wasp is proving effective and the cost of controlling the leaf hopper has fallen by 87 per cent.

The second technique that has proved successful in a number of cases is the 'sterile male' technique. Insects are sterilized, either chemically or by using radiation, and then released into the wild population to mate with fertile females who then fail to reproduce. If enough sterile males are used a marked reduction in the population soon shows itself. Radiation-induced sterility has proved successful in dealing with the screw-worm (a livestock pest) and the tropical fruit-fly. The International Atomic Energy Authority

(IAEA), the Food and Agriculture Organisation (FAO) and the World Health Organisation (WHO) are developing the 'sterile male' technique to try and eradicate the Mediterranean fruit-fly which destroys millions of pounds worth of fruit each year. Unfortunately, the sterile male technique using irradiation does not always work. Sterilized males of the anopheline species of mosquito, which carries malaria, are much less vigorous than the normal males and hence do not compete nearly so effectively as normal males for the available females.

Sterile males can also be produced by using the 'cytoplasmic incompatability' mechanism. This occurs when insects of the same species, but from very different areas of the world, are mated. The offspring, if any, are sterile. This technique was successfully used in a village in Burma in 1967 where *Culex pipiens fatigans*, an insect that carries the disease filariasis (the cause of elephantitis) was eliminated. The incompatible insect used was the product of insects from California and Sierra Leone. These sterile males are just as virile as the fertile ones and hence the test project was a resounding success. WHO and the Indian Government are now applying the technique in India in a seven-year campaign to eliminate the culicine mosquito.

WHO are obtaining encouraging results by using hybrid sterility. This means mating two different sub-species of the same species and producing fertile females and sterile males. Here again sterile males are as vigorous as the normal male.

A further method of control is to use chemical 'mouse traps'. Sex hormones, called pheromones, which attract male and female insects from long distances were the initial bait. They are difficult to extract and expensive to synthesise but are very effective in luring the insects to traps filled with pesticides. Dr Wright of the British Columbia Research Council has been experimenting with chemical substitutes that have the same seductive qualities. His preliminary trials with yellow jacket wasps have been successful. In the U.S. scientists have finally succeeded in synthesising the scent with which the male boll weevil attracts the female. Since the boll weevil causes $100 million of damage to the cotton crop each year, a safe method of eradication is badly needed. The great advantage of this system is that the insect comes to the poison and not the poison to the insect. This makes it effective against only one species, cuts down on the amount of poison used in absolute terms and avoids general pollution of the environment.

There are, of course, several drawbacks to these types of insect control. Biological methods work best on isolated populations of pests or those that are not too numerous. To try to eliminate the house-fly by releasing enormous numbers of sterile males into the environment would be not only a nearly impossible task, but also result in an unpopular plague of flies for some time. In addition to this, introducing the pests' natural predators into 'virgin' environments is an experiment that needs years of expensive prior research to evaluate the ecological consequences of releasing yet another alien body into the environment. One of the major factors that has to be checked is that complete elimination of an insect pest may cause a vacuum in the ecosystem that could be filled by an even more noxious species.

In 1959/60 the Chinese carried out an extermination campaign against sparrows. They were killed in their millions by poisoned traps, guns and keeping them flying so long by clashing cymbals that they fell to the earth exhausted and could be beaten to death. The unfortunate result of this highly successful operation was to unleash an unprecedented plague of locusts, normally kept down by the sparrows.

These biological methods of control, however, hold out great hope for the future. They have the advantage that they attack particular species rather than kill with crude indiscrimination like DDT and the organochlorines. Moreover, they do not add thousands of tons of dangerous chemicals to the environment.

At the same time, development of non-persistent chemicals specific to particular species is continuing. The most hopeful at the moment are the organophosphorous pesticides which, while more toxic than the organochlorines, are much less persistent and therefore less harmful in the long run — only a day after application crops are relatively safe. This process is slow, expensive and often proceeds by 'two steps forward and one step back'. Added to the continual need for new chemicals to overcome the development of immunity to old pesticides is the fact that very often, as with biological control, removing one source of danger merely releases another.

An analogous use of chemicals that are specific to a small range of pathogens is made against fungi. These are the systemic fungicides. They are termed systemic because they are absorbed from the soil water through the roots and spread throughout the plant in the vascular system. When a fungus comes into contact with the chemical, it dies. At present the results are quite promising. For they are both effective against the fungi and not seriously toxic to birds and mammals. Like insects, however, fungi that attack crops can breed fungicide-resistent strains. Already some new resistant strains of, for example, powdery mildew have evolved in America.

Another important method of keeping down the amounts of chemicals used in agriculture is to exploit the natural resistance of the plants themselves. Breeding new strains of plants is a lengthy process that demands infinite patience. A number of different varieties are chosen with the various desirable characteristics and crossed with each other in an attempt to combine them in an entirely new variety. Resistance to both pests and diseases is always a factor in these experiments. Two pests in rice are the green leaf hopper and the brown plant hopper. One variety of rice, Mudgo, is extremely resistant to hoppers while another, IR8, the new high-yielding strain which is spearheading the 'green revolution' in South-East Asia, is highly susceptible. The two have been crossed and preliminary trials have shown a plant both high-yielding and resistant. Similar efforts are being made against plant fungi, as for example with rust in wheat.

Always there is the danger, however, that beating one disease or pest will lay the plant open to greater attack from a different and previously unimportant pathogen. Perhaps for the future the most effective method will be a combination of both chemical and biological methods. Cereal mildews are a case in point. Plants are often susceptible as seedlings but resistant in the adult stage. A chemical seed dressing is used to control the disease to start with; intensive selection for adult resistance, it is hoped, will take over at a later stage of growth.

1969 was the year in which all these factors; increasing knowledge of the dangers of DDT, decreasing effectiveness as a pesticide and the growth of acceptable alternatives, all combined to stimulate a number of governments to ban or severely curtail its use. By 1970 most of the bans or restrictions announced by America, Canada, Britain, Sweden, Denmark, Norway, Australia and New Zealand were in force. Japan also banned the use, as from January 1970, of DDT and BHC on farms which raise livestock. Yet apart from Japan, the whole of Asia, together with all Africa, Latin America and much of

Europe are adding enormously to the world's residue of DDT every year.

It is essential that more countries follow suit, not only because of the harm that DDT does in the immediate vicinity, but also because, as we have seen, these harmful effects are not restricted but spread — and if anything, multiply — over vast areas of the world's surface.

One of the few ways that yet more countries can be persuaded to ban the use of DDT is to hit them financially through their exports. In 1967 a joint FAO/WHO committee fixed an 'acceptable' daily intake for DDT and other pesticides: 10.6 microgrammes per kilo body-weight for DDT and 0.1 microgrammes per kilo for dieldrin. This had immediate repercussions in Australia and New Zealand. New Zealand imposed various restrictions on the agricultural use of DDT in 1967 in order to protect her export market and Australia stopped the use of DDT in dairy pastures in 1968. This could prove a useful method of controlling the amount of DDT released internationally. For, if countries cannot sell their food because DDT levels are too high, they will be forced to cut down on the amount of DDT used.

The Wilson Committee, on whose advice the British Government acted, recommended in 1969 that the use of DDT, aldrin, endrin, dieldrin, TDE and heptachlor all be restricted. Dieldrin and aldrin had already been banned for a number of uses in 1964. Many scientists felt that the Committee did not go far enough in its recommendations, and advocated a complete ban.

There are several indications of the pressing need for a full world-wide embargo on the use of these persistent organochlorines: not just in a token smattering of countries round the globe which are, in most cases, those that were not using them in enormous amounts. Despite the fact that Great Britain now uses very little DDT, much still reaches it — borne on the Gulf Stream and the winds. Equally disturbing is the fact that scientists at Ohio University have calculated that over 2,000 tons of DDT and residues are locked up in the Antartic snow. Tests on Canada's polar bears in the Arctic showed that they had unexpectedly high levels of DDT. Both these places are thousands of miles away from where DDT is sprayed.

Nowhere is safe. There is as yet little evidence to suggest that DDT is directly harmful to human beings but, as we have seen, a correlation between high levels of DDT and deaths due to cancer, leukaemia and high blood pressure has been detected by scientists working at the Miami School of Medicine. The British Medical Journal has also reported a possible connection between impotence in farm workers and prolonged exposure to herbicides and pesticides.

At a more fundamental level still, these compounds may profoundly alter, and perhaps markedly destroy, the ecosystem on whose balance we depend. In such a case, instead of just a few isolated deaths from direct poisoning, we may have made the world uninhabitable.

While it is relatively simple for Western countries to ban the use of DDT and its fellows, it is very much more difficult for developing countries to take the same action. Even in Britain where there are no insect plagues, pests cause as much as £300 million loss of production per year and in developing countries the damage caused by pests to crops is so vast and the dangers of diseases like malaria so great that they cannot afford to stop using the organochlorine pesticides. The locust plague that threatened the whole of North Africa in the early months of 1969 was only averted by the intensive use of dieldrin sprays on the crops. Had the plague materialized, hundreds of thousands of people could have starved to death. Dr James Wright, chief of the vector biology and control section of WHO, has said that to limit the use of DDT in controlling malaria would lead to a 'major disaster'.

It is easy to see that the problem is extremely difficult if not totally intractable. It presents itself in the equation of 'our present certain needs versus our future survival'. It is less easy to see what the solution is, particularly since our knowledge is slender of just what form the long-term effects will take. It is only reasonable to suspect that the present evidence of menacing consequences is the tip of the iceberg. We must know more and that takes time and money. We must save lives now and for that we have the means and opportunity. But most important of all, we *must* plan for the future — something which takes more imagination and courage than the human race has often shown in the past.

VI

ATMOSPHERE IN PERIL

New York, London, Tokyo, Paris — most of the major cities of the world could disappear beneath the sea. Or the glaciers could once more move south and another ice age descend on the earth. It all depends on which form of air pollution — carbon dioxide or dust — gains the upper hand.

Temperatures on earth are regulated by the sun and the atmosphere. Radiation from the sun passes through the atmosphere, hits the earth and is re-radiated back into space as heat energy. Carbon dioxide in the atmosphere stops all the heat being re-radiated back into space because it absorbs it — acting very much like the glass in a greenhouse. Increased carbon dioxide levels mean more of the heat is trapped round the earth. For the past century the earth has been slowly warming up. This could lead to the polar ice caps melting and a rise in the sea level of 60 feet. All the low-lying lands of the world would be flooded. In the past 100 years alone the percentage of carbon dioxide in the air has increased by 15 per cent; this could rise to 25 per cent by 2000 AD if fossil fuels continue to be burnt at the present rate of increase. Carbon dioxide levels are now 330 ppm of air having been 290 ppm at the beginning of the century. At 600 ppm the earth's temperature could increase by 1.5°C.

Around 1960, however, the slow warm-up stopped and there has since been a slight reduction in the earth's temperature. In 1968 the Northern Atlantic ice coverage was the most extensive for 60 years. This is probably due to the other form of air pollution — the increased dustiness of the atmosphere. Since 1965 this has increased tenfold in many parts of the world. In the first place, this shields the earth from the sun's radiation. It also encourages cloud formation which reduces the amount of radiation received still further.

These particles of dust act as nuclei for rain clouds. The water vapour of which clouds consist, condenses round the dust nuclei and when the water droplets are large enough they fall as rain. Understanding this phenomena has allowed Man to encourage rainfall where it is needed by deliberately seeding clouds with small particles (silver iodide and salt, for example). But where air pollution does this inadvertently, the results can be most undesirable. From 1951 to 1965 the town of La Porte, Indiana, had 31 per cent more rain, 38 per cent more thunderstorms and 245 per cent more days with hail than neighbouring towns. The reason is that it has the misfortune to lie some 30 miles downwind of Gary and South Chicago where steel works belch into the sky vast plumes of smoke that act as nuclei for clouds to form round. La Porte is where the rain is deposited.

Jet aircraft have already increased cloud coverage over North America and Europe by five to ten per cent. Each time a jet crosses the Atlantic it consumes 35 tons of oxygen and produces about 70 tons of carbon dioxide, large amounts of water vapour and many fine particles. These form wispy cirrus clouds high in the earth's atmosphere. With the era of the giant supersonic jets already upon us, the cloud coverage could increase still further. The Nixon Administration has set a three-year deadline for the completion of a programme aimed at reducing aircraft pollution in the U.S. The new fuel burners which eliminate 70-80 per cent of solid particle emission should all be fitted by late 1972, at a cost of $13 to $15 million.

At present 31 per cent of the earth's surface is normally covered by clouds — it has been estimated that an increase of only five per cent could decrease temperatures sufficiently to cause a new ice age. The last ice age was caused by a temperature fall of only 7 to 9°C. The effects of carbon dioxide and the particles on the atmosphere and hence on the climate are immensely complex and not easy to prophesy. The normal functioning of the atmosphere is not yet fully understood — witness the frequent inaccuracy of weather reports — and the possible effects of pollutants on it even less so. It may be that a balance of pollution will be achieved whereby the warming effect of the carbon dioxide is cancelled out by the cooling effect of the particles. But there is no guarantee that this will happen and stringent regulation of all forms of air pollution is essential.

A major source of air pollution, especially in cities, is the car. Cars are responsible for about 60 per cent of urban air pollution in the U.S. — 86 million tons of pollutants, mainly carbon monoxide, in 1965. That year the atmosphere was contaminated by 142 million tons of pollutants, consisting of 72 million tons of carbon monoxide, 26 million tons of sulphur oxides, 19 million tons of hydrocarbons, 13 million tons of nitrogen and 12 million tons of solid particles. The problem is particularly serious in Los Angeles where seven million people using nearly four million cars in an area of 4,000 square miles cause the notorious Los Angeles smog. Car fumes contain carbon monoxide, nitrogen oxides, sulphur oxides, hydrocarbons — from unburnt petrol — and lead. The effect of sunlight on this noxious mixture produces the photo-chemical smog which causes damage to plants, animals, people and buildings. The pollutants, instead of being blown away, usually remain in the air above Los Angeles, giving it its characteristic greenish-yellow tinge. The geographical position of Los Angeles favours the formation of meteorological inversions — stable layers of cool air trapped by upper layers of warm air, thus concentrating the pollutants over one spot.

The hydrocarbons and nitrogen oxides react together under the influence of ultraviolet light and produce ozone. Ozone in excess, far from being the health-giver of popular imagination, damages the lungs and causes respiratory diseases. The position is such that when it reaches a concentration of 0.35 ppm, schools in the area are warned not to let children play energetic games. Since car exhaust controls were first introduced in Los Angeles in 1965, the hydrocarbon and carbon monoxide concentrations have decreased by 16 and 12 per cent respectively although the number of cars has risen by 11 per cent. However, the control equipment increases the amount of nitrogen oxide which is just as dangerous. Stricter controls are being introduced to curb this menace.

It was in California that the bill to ban the internal combustion engine by 1975 was introduced. The State Senate passed it but it was dropped by the Assembly. It probably served, however, to inform the car industries that concern over car pollution was very real. They are going to have to reach much higher standards of pollution control by the mid-1970s. In President Nixon's special message to Congress in February 1970, dealing with anti-pollution measures, he directed the Department of Health, Education and Welfare to establish new standards to control auto emissions in 1973 and 1975 models.

The U.S. already leads the world in its car exhaust control laws. France and Germany have limited laws; in England a recommendation that the carbon monoxide content of exhaust fumes be kept to 5.5 per cent in new cars was approved in February 1970, and in Canada exhaust pollution in Ontario is to be cut by 30 per cent in 1970, none of this approaches the American standards which aim to cut exhaust pollution by 80 per cent for the hydrocarbons and 66 per cent for carbon monoxide by 1975. California has the most stringent exhaust control laws in the world. Los Angeles, with a large, effective air pollution control agency with a staff of 300 and a budget in 1969 of $4.6 million, has few rivals for efficiency. While Federal law requires the car manufacturers to fit fume control equipment to new models, few States have done anything to ensure that the equipment remains effective once it leaves the factory. There have been cases of people deliberately disconnecting the control equipment to improve the performance of their cars. It is up to the individual States to apply the new Federal regulations and the results, so far, have been very patchy.

One proposal for cleaning up the car was to remove lead additives from petrol. Lead compounds were first added to petrol as an anti-knock agent. They raise the octane rating of the petrol, which allows a car to run more smoothly. Unfortunately it is also a poison and as little as 8 ppm in the body is toxic. In the U.S. about 250,000 tons of lead are discharged into the air each year; in Britain about 3,000 tons. Although there has been little evidence to suggest that the amount of lead released is hazardous (the average American body contains 2.5 to 3 ppm) it is unwise to allow the background levels of this poison to build up. It already pollutes the air over the oceans, and roadside plants contain high concentrations. Car manufacturers will have to fit some sort of catalytic afterburner to the exhaust system to reduce unburnt hydrocarbon emission. Since the lead compounds in the exhaust gases poison the catalyst unacceptably fast, the lead must be removed.

The cars manufactured by the General Motors Corporation in the U.S. in 1971 are testing whether the public is really sincere in its desire to avoid polluting the air. They have been modified to run on unleaded petrol. This has meant a loss in performance and economy. General Motors calculate that a typical big-engine car will lose one-tenth to one-half a mile per gallon in fuel economy and one-tenth to one-half a second in acceleration. There will also be an eight per cent horse power drop using the less powerful petrol. Avoiding pollution is expensive.

Because of the increased interest in pollution-free cars, experiments with electric and steam cars, which might have performance without pollution, are taking place. The first International Electric Vehicle Symposium was held in Phoenix, Arizona, in November 1969. A number of different electric cars were demonstrated. At present one of the main difficulties in producing electric cars is the battery needed. It occupies too much space in the car — the whole boot — and needs to be recharged after the car has cruised for 80 miles. By the time electric cars are ready to go into production the pollution problems of the internal combustion engine may well be cured.

The National Air Pollution Control Administration in America has started laying down air quality control standards for industry and other stationary sources of pollution. But implementation of these standards in the individual States has been somewhat erratic. Industry still wields a great influence over the State legislatures and many of the regulating boards are packed with industries' spokesmen who block effective action on pollution. In Texas, lobbyists for the State's 1,300 cotton gins succeeded in exempting them from controlling their emissions in 1967 and it was only in 1969 that, after a two-year struggle, the exemption was nullified. Air pollution control equipment is expensive and industry is reluctant to spend money. On many occasions plants have closed down rather than install the necessary equipment.

Texas is one State where the air pollution controls are now being seriously enforced. In 1969 action by the Texas Air Control Board caused the closure of at least four plants, including two lumber mills, that did not reach the State standards. A chemical company was forced to close in 1970 because it had not controlled fluoride emissions that were described as 'strong enough to etch glass'.

Fluorine is one of the most serious pollutants produced by industry. It is discharged by, among others, brick works, aluminium smelters and steel plants. Although fluorine is emitted only in very small quantities, once it falls to the earth it is concentrated in the grass. Cattle then eat the contaminated grass and are poisoned. The symptoms of fluorosis are softened bones and loss of teeth. Fluorine also damages fruit and crops. Control of fluorine pollution is expensive and is costing millions at the new aluminium smelters being constructed in

Anglesey and Northumberland in England. The usual solution is to build very high chimneys and hope that the winds dilute the fluorine to harmless levels before it is deposited.

Few U.S. States have done anything about banning open burning and backyard incinerators. These are a significant cause of air pollution in cities and are also considered outdated as a form of waste disposal. Many municipal incinerators cause considerable pollution and people living downwind of them get their homes and gardens covered with a fine layer of black ash. The reason is that many incinerators, being old and inefficient, do not have anti-pollution devices and are not able to deal with the increasing volume of garbage. This is fed in too fast and the incinerators are not hot enough to burn the garbage completely. Black soot and smoke spew forth, contaminating the air.

Nevertheless, despite opposition from industry and dilatory action by many authorities, there have been some encouraging successes in dealing with air pollution. The sulphur dioxide concentrations in New York's air are falling though it is still the most polluted city in the U.S. It is down by 56 per cent compared with the 1966 levels. This has been due mainly to a shift from high-sulphur oil and coal to low-sulphur varieties. Use of fuel with more than one per cent sulphur was banned in October 1969. Particulate emissions — a measure of soot — have been cut by 23 per cent since 1966. This reduction was due partly to low-sulphur fuels having a low ash content and partly to the shutting down of some inefficient municipal incinerators.

In England since the Clean Air Act of 1956, there has been a marked improvement in the atmosphere. The amount of smoke in London's atmosphere has been reduced by at least three-quarters — there will be no repetition of the notorious smog of 1952 that was responsible for the deaths of 4,000 people. But despite the dramatic improvement in the quality of London's air there are still some parts of the country where the Clean Air Act is not being properly applied and 'dirty' coal is still being burnt. This is mainly in the North where many local authorities have clean air schemes that will not be completed until the mid-1980s. This dilatoriness on the part of local authorities is a strong argument in favour of making the clean air measures compulsory on a nation-wide basis.

The Clean Air Act has meant that the use of oil and natural gas has steadily increased. These do not produce the smoke of the coals but much of the oil and some of the gas has a high sulphur content, which the Clean Air Act did not cover. Some seven million tons are discharged into the atmosphere in England each year. Although sulphur dioxide pollution in Britain itself has not increased much in the past three years, this is mainly due to the fact that the factories have higher chimneys and since Great Britain is a long, thin island the winds can blow the sulphur dioxide away before it has a chance to land.

ATMOSPHERE IN PERIL

Unfortunately it has to land somewhere, and Scandinavia appears to be the favoured recipient. The sulphur dioxide mixes with the rain to form sulphuric acid and falls on the land. The Scandinavians claim that growth in the forests is affected and the lobster and fish populations are damaged. They blame the industrial areas of Holland, Germany and England for this pollution.

One of the difficulties of dealing with air pollution is that it crosses international boundaries and it is often impossible to identify the source of the pollution. The first fully automated early earning system in Europe went into operation in Rotterdam, Holland in early 1970. It is known as the Philips SO2 system and is controlled from a central office called STENCH control. Monitors have been installed to cover the Rijnmond area of Rotterdam. They automatically record the sulphur dioxide (SO2) concentrations of the air and relay the information to a computer at STENCH control. When the sulphur dioxide rises above a certain level, an alarm sounds and the operators in STENCH can pinpoint the source of the pollution. The culprit or culprits are then asked to control their emissions. This system can be extended indefinitely. A network covering Europe could do much to identify offending industries at source and hence facilitate the task of stopping them.

Sweden is itself reducing the amount of sulphur dioxide it produces. The Government is providing grants to industry to install sulphur dioxide purification plants, and any new industrial plant has to prove that it will not contaminate the environment before planning permission is given. The Government has also encouraged the changeover from heavy to light oil in central heating, since it contains much less sulphur. The amount of sulphur dioxide in the air could be drastically reduced if sulphur-free fuels were burnt, even though these do cost more. Paris insists that only sulphur-free fuels be used — after the massive clean-up of its national monuments in recent year, it would be stupid to allow them to become blackened again. The sulphuric acid attacks the surface of the stone and slowly erodes it — the black covering of soot that covered the monuments had one good effect at least, in that it protected them from being eaten away. Once the sulphur dioxide and soot content of the air in cities is permanently reduced, it becomes worthwhile to restore the buildings to their pristine glory. In London work has already started on cleaning-up Buckingham Palace and a number of other famous buildings.

The damage to buildings, while unsightly, is not the most serious effect of urban air pollution. The complex mixture of air pollutants in all cities presents many known and some unknown health hazards. There are traces of known carcinogens in the atmosphere which may be encouraging cancer growth and environmental factors may well be a cause of congenital defects in babies. The respiratory diseases associated with smog are well-known, in fact the incidence of deaths due to bronchitis has been used as a measure of air pollution. In Europe, Britain heads the list with 28,257 deaths a year, followed by West Germany with 10,000, France with 2,606 and Holland with 1,768. The photo-chemical smogs that are common in America affect the eyes as well as the lungs — dark glasses are worn in New York to protect the eyes not just as an affectation. These smogs are the result of the effect of sunlight on car exhausts and are not likely to affect England where there is not enough sun. Nevertheless carbon monoxide levels in many London streets are dangerously high and sufficient to induce unpleasant headaches in those exposed to them for very long.

In the world's most polluted city, Tokyo, traffic police have to wear masks and report back to headquarters frequently for a revivifying whiff of oxygen. Their lungs are invariably cleared when they go off duty. Tokyo registers the highest pollution levels in the world with both industries and cars contributing. Athletes going for an early morning run in this atmosphere are endangering their lives. Fresh air can be found in Tokyo — at a price — a shilling a time from oxygen-vending machines. The Government is slowly taking action — sulphur dioxide discharge limits have been set and an air pollution control centre planned. The centre will be connected to each factory that discharges sulphur dioxide by a 'hot line'. It will check the amount of gases being emitted and warn the factory when it reaches danger level. However, no car exhaust controls have yet been introduced.

POLLUTION: THE WORLD CRISIS

Ankara, capital of Turkey, is a close rival of Tokyo in winter. There it is not the fault of the motor car but of coal. Soft coal with a high sulphur content is burned to provide heat. This blankets the city with a thick, black pall of smoke that sometimes reduced visability to ten yards. A 1969 study by the Ankara Medical School found that pollution in the city had increased 42 per cent since 1965. At the same time lung cancer has increased by 9½ times and chronic bronchitis 11 times since 1962.

A major air pollutant, carbon monoxide, is well-known as a killer in acute amounts since it combines with the haemoglobin in the blood to form carboxyhaemoglobin which is incapable of carrying oxygen and the body suffocates. It is its chronic effects that are most disturbing, especially to cigarette smokers who inhale it. Recent experiments in Denmark have shown that it could be an important cause of heart and arterial disease. Three scientists from the Rigshospitalet in Copenhagen, Drs Kjelsden, Wanstrup and Astrup, subjected rabbits to a combination of small quantities of cholesterols (the fatty acids found in butter and milk) and low concentrations of carbon monoxide (no more than is normally found in the blood of a fairly heavy smoker). After ten weeks these rabbits had two and a half times more fatty deposit in a vital artery than those who had no carbon monoxide. As Prof P.J. Lauther, director of the Medical Research Council's Air Pollution Research Unit at St. Bartholomew's

Hospital, London, commented, 'This is another nail in the smoker's coffin.'

It has also been shown that high levels of polluted air impair driving ability. Experiments carried out by Sussex University and Brighton College of Technology, published in *Nature* in January 1970, showed that breathing polluted air decreased mental efficiency. The tests were carried out with the subjects breathing the polluted air from a fairly busy road and then breathing pure air from cylinders. They performed the tests worse breathing the contaminated air which, by central London standards, was not heavily polluted. Carbon monoxide levels in Oxford Street sometimes reach 360ppm. In New York, studies have indicated that exposure to 50ppm of carbon monoxide can affect the mental processes drastically. This data indicates that air pollution may play a hitherto unsuspected role in road accidents.

Air pollution is expensive. In England the annual bill from lost production due to illness, damage to buildings etc. is estimated at some £350 million. In the Los Angeles Basin, once a fruitful area, lettuce and spinach are no longer grown, citrus fruit yields have been halved and horticulturalists have had to move out of the area. A group of American scientists reported in January 1970 that chlorotic dwarf disease among white pine trees was caused by a combination of sulphur dioxide and ozone pollution. The disease causes the pine needles to turn yellow and they are shed prematurely; it eventually stunts the growth of the entire tree. This has caused an immense amount of damage in California alone, where more than a million trees were reported dead or dying in December 1969. It is the *combination* of the two gases (typical constituents of the Los Angeles smog) that appears to do the damage — individually they are harmless.

[Conclusion] Air pollution is an international problem that requires international co-operation to deal with it. It is little use countries such as Sweden acting to curb their own pollution when the winds — which are no respecters of frontiers — carry sulphur dioxide from other countries and deposit it on their land. There has been a steady increase in the acidity of rain all over Europe during the 1960s. The 'black snow' that fell on eastern Norway and western Sweden in 1969 came from the Ruhr in West Germany. It was, in fact, greyish with black spots and had a high sulphuric acid content. The Norwegians and Swedes fear that if sulphur dioxide continues to increase, growth in the forests could be severely inhibited which would lead to heavy economic losses. Sulphuric acid dissolves essential minerals out of the soil which are drained into the rivers and lakes. This not only impoverishes the soil but also upsets the ecological balance between plant and animal life in the rivers. It has been calculated that raising the acidity of a soil can reduce growth by as much as five per cent a year.

Many forms of air pollution are already being dealt

with. Smokeless coal is being produced which cuts down the amount of soot in the atmosphere. Industry has switched from coal to oil and natural gas as a main energy source. Sulphur-free oils are being produced which, while adding to the cost of heating, will cut down the cost of air pollution. Pollution from the internal combustion engine can and is slowly being controlled. Stricter standards are being introduced to deal with industrial pollution and these are being gradually enforced. The growing use of nuclear power should mean that the amount of carbon dioxide and particulate matter — which have such a potentially catastrophic effect — that is discharged into the atmosphere is reduced. Unfortunately nuclear power itself is potentially very dangerous (see Chapter VII).

Further research into the potential side-effects of global pollutants such as carbon dioxide, particulates and sulphur dioxide must be carried out. One atmospheric mystery as yet unsolved is where the 200 million tons of carbon monoxide produced each year go to. The background amount of carbon monoxide has remained constant at about 0.1ppm despite the vast amount that has been pumped into the air.

Air pollution can be controlled, but only by increasing the costs of manufacturers. International industry is competitive and it may only be by international agreements that this massive threat to a stable climate, clean air and a healthy population will be countered.

VII

NUCLEAR POLLUTION

Nuclear power is here to stay. By 2000 AD, as much as 40 per cent of the world's electricity will be supplied by nuclear power stations. Nuclear explosions will be used in engineering projects and nuclear weapons may continue to be tested underground. Nuclear power is a cheap and relatively inexhaustible source of energy.

With so many advantages, this rapid increase is, on the surface at least, entirely justified: just another tool that man has developed in his struggle for survival.

But, parallel with this increased use, comes the increased risk, on the one hand, of a major disaster such as the explosion of a power station and, on the other, the chronic, general contamination of our environment.

The horrors of a nuclear explosion are well-known from Hiroshima and Nagasaki (1945): a holocaust of thousands, and for the so-called survivors a life of fever, nausea, loss of appetite, anaemia, vomiting, diarrhoea and finally coma and death. Exposure to radiation induces a weakening of the whole body which makes it extremely susceptible to attack by any disease especially cancer, leukaemia and congenital deformities in children. Even a cold can last six months.

NUCLEAR POLLUTION

It is the effects of a general contamination, however, that are perhaps more threatening in a world that may preserve nuclear peace. As is true of so many aspects of the whole pollution crisis, the effects are only just being studied in depth. In the first years of the nuclear era the possible seriousness of having low but constant levels of radiation in the world around us was ignored and it was only lethally high doses that were considered as important. (The same problem was seen with DDT.) Today, however, there is a slowly accumulating body of evidence which suggests that even at relatively low levels it can cause irreparable damage, particularly to children.

Radioactive material, like strontium-90, iodine-131 and caesium-137, passes readily through the metabolic processes of the body, radically affecting the nuclei which are the centre of cells and vital for cell multiplication. Substances such as strontium-90 can kill cells outright: strontium is stored in bone marrow and produces leukaemia. Radioactive material such as tritium, for example, may derange the process of nuclear division and so produce malformed children.

An extreme form of this was revealed in a report by the Atomic Bomb Casualty Commission which examined mentally retarded children affected by the Hiroshima and Nagasaki bombs. It was found that with an increasing exposure to radiation there was an increasing incidence of mental abnormalities. The children of women who were pregnant at the time of the bombs in 1945 were studied and the results are startling. As many as 36 per cent of children exposed to 200 rads and upwards (a rad is a unit measure of radiation) were mentally retarded; 9.3 per cent with an exposure of 100 to 200 rads and 4.55 per cent with 50 and 100 rads. Perspective is given by a control group of 500 women who were away from the cities at the time: their children showed less than a one per cent abnormality rate. Further signs of the harmful effects of the mothers' exposure to fallout were a general tendency for the children to have small heads and an incidence of malignant tumours of the central nervous system higher than the control group by as much as 40 per cent.

Painstaking work has been carried out on the effects of a general and constant level of radioactivity. In 1959 Dr John T. Gentry published the results of a survey he carried out in New York State. He selected three areas which had unusually high natural levels of radioactivity because of the particular rock types they stood on, and compared the birth certificates of children born over an eight year period (1947-55). He found conclusively that in those three areas the incidence of congenital malformations in children was always higher than in the rest of the State. He also found that the risks were greater in rural areas than urban ones, which are more sheltered from the radiation-emitting rocks.

Far greater controversy was aroused by the work of Professor Ernest Sternglass of the University of Pittsburgh. In 1969, he claimed that over 400,000

NUCLEAR POLLUTION

children in the U.S. had been killed by hitherto unsuspected side-effects of strontium-90 fallout from the nuclear tests carried out between 1945 and 1962. Strontium-90 in *high* levels is a known hazard. It is absorbed by grass and when eaten by cows turns up in the milk that children drink. The body treats it in the same way as calcium — it stores it in the bones. Many years may pass before its pernicious effect is revealed, but it leads to leukaemia and bone-marrow cancer. It was generally believed that the strontium-90 level to which the population had been exposed was not sufficient to cause genetic damage. Professor Sternglass has now claimed that it was — and is. He believes that large numbers of children have died either in the womb or shortly after birth as a result of a genetic effect of strontium-90 concentration in the milk and bone.

The infant death rate declined steadily since 1934 until the early 1950s when the rate of decline slowed down. Professor Sternglass believes that had there been no nuclear tests infant deaths would have continued to decrease at the same rate as during the 1934-50 period. He extrapolates the curve of the 1934-50 graph and treats the difference between it and the actual graph as the excess deaths caused by strontium-90. He relates the postulated increased mortality to the increased strontium-90 fallout, showing that this excess mortality varied with time in exactly the same way as the strontium-90 content in the milk varied. He has estimated that 400,000 babies in the U.S. and 100,000 in Britain died because of

POLLUTION: THE WORLD CRISIS

this effect. His claims are hotly disputed by most scientists.

A point by point rebuttal of his hypothesis by Professor Rotblat and Dr Lindop of St. Bartholomew's Hospital, London, was published in *Nature* in 1969. The main point of dissension was the validity of Professor Sternglass' extrapolation of the 1934-50 rate of decline. While the curve certainly started levelling off in 1950, many other factors that Professor Sternglass did not consider could have caused this. During the twentieth century infant mortality has steadily declined with increasing standards of living and medical care. The introduction of the sulphonamides and other antibiotics all increased the numbers of children that survived the first years of birth. The impact of these new drugs in infant deaths was initially great, but once the diseases that they cure are eliminated, the rate of decline, though sharp to start with, decreases (until another cause of infant deaths is found and eliminated). Thus extrapolation from the steeper points of the curve is not valid. As diseases are eliminated, infant mortality rates will continue to decline but no part of this rate of decline can be treated as the normal, expected rate. A levelling out of this rate does not necessarily imply that some external factor is raising the death rate.

There are now two major sources of artificial nuclear contamination which can cause the sort of effects discussed. They are the explosion of nuclear

NUCLEAR POLLUTION

devices for both military and civil purposes and the use of nuclear reactors to produce electrical power. The development of nuclear-powered motors for submarines, ships and cars poses a possible threat for the future but for the moment it is the first two that are the most dangerous.

The centre of a nuclear power station is the reactor which is, in effect, a controlled atomic bomb in a constant state of almost-exploding. It does not explode because it is not allowed to get too hot. This control is achieved by circulating water round the reactor. The resulting steam turns the electrical generator and is then passed through a cooling tower and recycled to the reactor. The water becomes increasingly contaminated by radioactive matter that leaks from the reactors. When it is withdrawn from the cycle, it has to be purified and the pure but hot water is returned to the rivers causing thermal pollution. The radioactive wastes then have to be disposed of. At the moment they are either stored in underground caves in steel containers or dumped in barrels in the sea.

An even more difficult problem is the disposal of the waste from the reactor itself. Each ton of uranium processed produces 100 gallons of extremely hot and concentrated radioactive material. In the U.S. the Atomic Energy Commission (AEC) has already stored over 80 million gallons of intensely radioactive materials in nearly 200 underground tanks. This was the waste of the nuclear weapons programme. It is estimated that 3.5 million gallons of high-level liquid

NUCLEAR POLLUTION

waste from the civil nuclear power programme alone will be produced by 1980 and 60 million by the year 2000. These liquids will still be dangerously radioactive after hundreds of thousands of years. Millenia after this civilisation has passed away, it will still be a potential danger to earth.

These tanks have to be artificially cooled to prevent the intense pressure generated by the boiling liquids from bursting them. An estimated 55 million gallons is stored at Hanford in Washington State, in an earthquake-prone area. The reactors there produced the plutonium for the bomb that destroyed Hiroshima and Nagasaki. During their operational years they produced enough plutonium to destroy the whole world. The radioactive wastes were then stored in 140 tanks buried underneath the earth. Twenty years after the storage site had been selected, a geological survey of the area was made for Douglas United Nuclear who had taken over operation of the reactors for the AEC. The survey revealed that the earthquake hazard of the site was even greater than had been initially thought and the tanks were not designed to withstand any nearby earthquakes. They have to survive for hundreds of thousands of years — for the radioactivity stored in any one tank could be as much as all the fallout the world has yet produced. Even if the tanks were not destroyed by an earthquake, the cooling systems might be damaged which would be just as hazardous. Hot radioactive gases would be released into the atmosphere spreading havoc in their path. Liquid wastes from

NUCLEAR POLLUTION

broken tanks would seep though the earth, contaminating the ground water, eventually poisoning the Columbia River and the Pacific. The tanks are also an obvious target for an enemy bent on releasing radioactive material into the U.S.

In their book, *Perils of the Peaceful Atom*, Richard Curtis and Elizabeth Hogan pointed out that some of the storage tanks had failed after only 20 years. The complex cooling apparatus that is needed to keep the tanks from exploding require constant surveillance. As a result of these problems the AEC are experimenting with new methods of dealing with the wastes. When solidified they can be stored in one-tenth of the volume needed for the liquid radioactive materials, and be disposed of more easily. The Waste Solidification Engineering Prototype started operating at Hanford in November 1966. However, the process is slow. By June 1969 less than one per cent of the total wastes at Hanford had been processed. These solid wastes, which have low levels of radiation, have been dumped in containers in the sea and stored in barrels underground. The West Germans are storing their low-level wastes in salt mines over 2,000 feet below the ground.

The Americans have developed a 'hydraulic fracturing' technique whereby intermediate-level wastes are injected into the ground. The wastes are mixed with cement and pumped under high pressure into fractures in shale formations 700-1,000 feet down. The mixture forms a thin, hard sheet. This technique is being used at the Western New York

POLLUTION: THE WORLD CRISIS

Nuclear Service Centre at West Valley, New York. It is being monitored closely to see how well it works.

The sea has been used as a dumping ground for radioactive wastes on the principle that even if the containers leak, the radioactivity will have died down to a safe enough level to be diluted. Since the high-level wastes remain lethal for hundreds of thousands of years, because many of the radioactive isotopes within them have very long lives, the containers should be virtually everlasting, which is by no means true. In any case, the capacity of the seas to dilute the wastes has been overestimated. Both plankton and fish readily absorb radioactive phosphorous and strontium-90. Plankton contain 50,000 times as much strontium-90 as the seas around them and fish eggs 100,000 times as much. This concentration of radioactivity can poison the man who eats the fish. One thousand rads are fatal to Man and 50 rads causes genetic damage and induces cancer.

There is a further source of radioactive pollution from power stations apart from that caused by waste disposal and that is the day to day operation of the plant itself. While the safety standards imposed on the nuclear industry are higher than those in any other, some leaks of radioactive materials do occur and there is always the danger of a major accident. These dangers will increase with the growing number of nuclear power stations. Should there be a failure, for instance, in the water supply to cool the reactor, it would explode. With a growing number of power stations being sited nearer and nearer major population centres, the possibility of a nightmare disaster increases. In 1970 a study group for the Colorado Committee for Environmental Information reported that the plant at Rocky Flats had been leaking dangerously radioactive plutonium into the environment for years. Plutonium is about the most lethal material ever produced by Man. It emits hard alpha radiation which destroys the cells of the body and when ingested — from the air, water or in food — is lethal. Dr Edward Martell, who led the study group, reported that plutonium from the plant could be found four miles away from its source. The ground sampling and monitoring programme at Rocky Flats is now to be extended. It is possible that stringent though the regulations are, they are still not safe enough. Some scientists believe that there is no radiation dose that can be called safe.

The Rocky Flats area is doubly unfortunate because it is also the site for the production of nerve gases. The disposal of the wastes from the process has provoked earthquakes and threatens the water supplies of the region. In 1960 the U.S. Army buried the wastes in a 12,000 foot well near Denver. In 1962 Denver had its first earthquake and since then there have been over a thousand minor ones. The waste liquid that was pumped into the well acted as a lubricant on the rocks below Denver. They began to slip, one over another, and each time this happens there is another earthquake. Apart from this hazard, there is also the danger that these toxic wastes could

POLLUTION: THE WORLD CRISIS

seep into the water supplies and turn up in the city drinking water. Liquids can run along the fractures in rocks caused by the earthquakes and eventually contaminate the rivers and underground water table.

The second source of general contamination of Man's environment is the explosion of nuclear devices. Initially, of course, these were all for military purposes and carried out in the atmosphere, allowing all the radioactive material to spread into the atmosphere. Since the Nuclear Test Ban Treaty of 1963, however, all testing has gone underground: except for the French and Chinese who have not subscribed to the ban. This has been paralleled by the growth of civil projects, using underground explosions, in the American Plowshare programme which was first set up in 1957. It aims to use nuclear power for engineering purposes — mining, clearing harbours excavating canals, etc.

The fact that these explosions are invisible has produced a sense of security more imagined than real. Any hope that they are entirely safe underground is unfounded since there are still leakages into the atmosphere. The stresses set up in the earth's crust in the area of the explosion often leads to faults in the surface rock being opened up. Radioactive materials leak out into the environment through these faults. While the degree of fallout is a fraction of that produced by the atmospheric tests, it is still a dangerous and continuing source of radiation.

In the Plowshare programme there are considerable

risks of a more direct contamination because of the unpredictable reaction of the rock above the blast. One, called Sulky, was 90 feet down and produced only a depression in the ground. Palanquin on the other hand, which was set off 280 feet down, erupted through the surface, sending up a great ball of fire. A large proportion of the radioactive material produced was released into the atmosphere. Even 'clean' Plowshare devices with 99 per cent fusion and only 1 per cent fisson are dangerous because they produce massive amounts of tritium which can combine with the cell nucleus and cause permanent chromosome damage. In any case, Plowshare scientists have calculated that at least 10 per cent of radiation material produced by blasts during the course of the programme will escape unavoidably.

One of the projects in Plowshare is the construction of a new canal across Central America to replace the Panama. Its likely effects have been graphically described by Dr Martell of the National Centre for Atmospheric Research. 'The ejecta lip will form a thick, unsightly layer of radioactive mud and rock in a swathe several times as wide as the canal. Throwouts and air blasts will extend the devastation by flattening forests and structures for miles in each direction. Seismic and acoustic waves generated by the nuclear blasts will produce unpredictable levels of damage up to distances of tens to even hundreds of miles. And there will be a serious concentration of some radionuclides in the terrestial and marine biosphere in nearby, downwind and downstream areas.'

One side-effect of many of these tests is the earthquake hazard that Dr Martell mentioned above. Most of the underground tests have taken place in isolated areas which are often geologically unstable. The enormous power of the explosions in such conditions has been enough to trigger off an earthquake. For example, testing in Nevada triggered off an increase in seismic activity in the area that lasted for some days.

When the bombs are used for mining, there is not only the danger of radiation leaking and earthquakes being set off but also the desired products may be contaminated. A device in the Project Gasbuggy programme, which is part of Plowshare intended to enlarge natural underground gas reservoirs, was exploded in 1967. It was found that although radioactive iodine was not a hazard, both tritium and radioactive krypton were contaminating the gas. It was also found that the gas taken initially from the well had been diluted to half its volume by a mixture of carbon dioxide and hydrogen, presumably produced by the explosion. The hydrocarbon content of the gas has now reached 90 per cent. However, the radioactive elements released in the explosion would have to be removed before the natural gas could be distributed to homes. It is at least 25 times as heavily laden with radioactivity as can be safely used and so has to be burnt on the spot. With all these disadvantages — general nuclear pollution of the environment, earthquakes, and contaminated end-products — this 'peaceful' use of the nuclear power for civil projects seems a tool Man could well do without.

In the midst of all the controversy about the hazards of general nuclear contamination of the environment, two conflicting points should be borne in mind. One is that whether or not Dr Sternglass — and the other scientists who, like him, but not in such extreme form, have protested that radiation even at low levels is harmful — are right or wrong, it is important that no possible cause of damage to Man be overlooked. If nothing else, Dr Sternglass' work should stimulate a very rigorous search for any unsuspected side-effects of radiation. Although nuclear fallout had added very little radiation to the natural background of 0.1 rads, even this infinitesimal amount, 0.0013 rads a year (plus 0.0003 to 0.003 rads from waste disposal), is potentially harmful. The second is the danger of 'crying wolf' too often. The present regulations controlling radiation doses are stringently applied. Public indignation raised unnecessarily could in time lead to indifference towards these standards.

Because there are not enough coal, oil and natural gas reserves in the earth to meet the power demands of future generations, nuclear power is an essential part of our future development. But the strictest precautions must be taken to ensure that the radiation level of the environment does not increase any further, and this means not only that nuclear explosions should not be allowed to contaminate the atmosphere but also that completely safe methods of disposing of lethal wastes must be found. The sea has already received as much waste as it can safely tolerate and inefficient nuclear storage tanks on land are a permanent hazard. It has been suggested that the longer-lived radioactive elements could be concentrated and shot into space where they could decay until they are harmless. This is not yet a practical solution but it may be used in the future. An enormous, lethal garbage dump floating endlessly in space. . . hardly a prepossessing thought but typical of the arrogance of Man's attitude to the world and universe around him.

VIII

TOO MANY TAME PEOPLE

Cities used to be garb, now they are garbage
                          Marshall McCluhan

There is a theory, well-documented in the animal kingdom, called 'social stress'. It is a phenomenon of sudden, abnormal behaviour patterns, a society running amok, killing itself. This, the argument runs, is the result of over-population and is triggered off not, as one would expect, by too virulent a competition for available food but rather by the overwhelming strain of having too many individuals pressing in on each other.

The most famous example of this social stress is the mass suicide of the lemming. Every three to five years, hundreds of thousands of lemmings in Scandinavia swarm down from their regular breeding grounds. In the 1970 migration these normally extremely timid animals attacked cats, dogs and even humans in their journey. When they reach the coast, they throw themselves into the sea; their bodies litter the shore for months. That this is not a rather vicious Malthusian method of birth control made imperative by a lack of food is shown by the fact that it occurs in spring when there is plenty of food about.

The same sort of problem has been discovered in the snowshoe hare. R.G. Green has for many years been examining them round Lake Alexandra in Minnesota, U.S. He found that every ten years there was a population crash in spring. Suddenly large numbers died for no apparent reason: they just gave up. Even some he caught at the time and kept in captivity died. Again, no question of shortage of food.

In the laboratory, rats have been subjected to conditions of overcrowding, noise and light. The results are frightening. All their normal behaviour patterns break down. They become diseased and start fighting each other. They eat their young and even mutilate themselves. This continues until so many have died that the population is reduced to a manageable level again. Normality returns. All these cases share similar patterns of behaviour, this extraordinary aggressiveness and this blind death-wish. This catastrophic destruction of population is caused by too much contact, too many stimuli — the natural end-product of over-population.

The implications of social stress for Man are obvious and, to say the least, unpleasant. The population is growing fast and the necessity of providing for that population compels an ever increasing percentage to live in towns and cities. There is just not enough land available both to house people in open, natural surroundings and feed them adequately. Since it is easier to concentrate people rather than food production in a small space, it is people who are compressed and crammed. The result is that cities spread into conurbations. An environment of exciting, invigorating interaction (which has generated most of Man's achievements) changes into one of constant aggravation and enervating struggle. It is a development that Patrick Geddes, the great Scottish town planner of the early part of this century, who first coined the phrase 'conurbation', defined as 'slum, semi-slum, super-slum'.

The question that immediately poses itself is whether the social stress seen in animals as a result of over-contact is likely to appear in Man, when he is pushed into a similar situation in the modern city. Could, for example, rising levels of suicide, drug addiction and homosexuality all be facets of the problem? Could student demonstrations, industrial strikes and rising social frustrations — evidenced all over the world — be in some way or other attributable to social stress? If they are, or there are other symptoms in evidence, then it is obvious that immediate steps must be taken before Man rushes like a lemming to his death.

Lewis Mumford, in his *City in History*, asserts that the ideal size for a city is that of the Greeks or Romans, which combines both easy access to the countryside and what Robert Ardrey in *The Territorial Imperative* calls the 'noyau' or nucleus effect of strong, challenging social contracts. Today it is obviously impossible to resettle the people in cities of that size, but dividing them into so-called 'precincts' centred round, say, trafficless shopping areas, might go some way towards creating the same effect. This would bring back human scale to cities: it would encourage involvement and participation in the life of the city.

Some sort of control of the car is essential if cities are to remain habitable. Too often towns and cities are carved up for the convenience of the car: it becomes easier and easier to get somewhere that is less and less desirable to reach. There is, however, the curious example of the model town of Glenrothes in Scotland which was planned on the best principles of traffic-free pedestrian zones. The inhabitants find it unexciting and desert the town centre for that of nearby Glasgow with all its disadvantages.

It is not necessarily the urban slum environment that is the most damaging to Man. The planned, modern housing estates of the developed countries that are efficient and sanitary have a more insidious effect. Man has been so successful because he is the most adaptable of all animals. He can live as a carnivore or vegetarian, he can live in town or country under any system of government, he has adapted himself to life in the desert, jungle, temperate and arctic environments. But always his genetic inheritance has been formed by living as part of and with nature. Many modern conurbations do nothing to satisfy Man's biological and emotional needs. Young children need many different stimuli in the first years of life to bring out their potential intelligence — if they live in a featureless environment their intellectual and mental growth will be stunted. For these reasons alone, recreational areas, game reserves and parklands are all necessary for Man's continued healthy development.

Man can cope with the overcrowding in cities by adapting his behaviour — but only up to a point. The phenomenon of violence for its own sake — exemplified by street gangs, 'skinheads', 'rockers' and so on — are symptons of breakdown. Indeed, the whole syndrome of disaffected youth may be to a significant degree traceable to the conforming pressures and de-humanizing influences of city life. The biological needs of Man are in the long run more important than his need for shelter and comfort.

Urban Man is, of course, heir to many ills apart from overcrowding: air pollution and noise being the chief. High noise levels impair hearing and lead to inefficiency and carelessness at work. According to a report published in *The Times* in March 1970, this probably costs between £1,000 and £2,000 million a year. There are, admittedly, strict regulations governing aircraft, car and motor-bicycle noise (the two latter are confined to 89 decibels (dB), but all too often they are not enforced. Problems of measuring the sound level and lack of men on the spot have prevented the police from using their potential powers. In France, on the other hand, where the level is fixed at 83 dB repeated offenders have their cars confiscated. In Switzerland the level is even lower at 70 dB during the day and 60 dB at night. Residential areas in many Japanese cities have a limit of 50 dB — and these regulations are strictly enforced.

POLLUTION: THE WORLD CRISIS

There are ways, then, that the urban environment can be improved: clean air, less noise, and precincts which would be easy to identify with. But further than this the country outside the towns must be improved to provide the vital contact with nature on which a balanced outlook on life depend. The urgency of the problem is made clear by the fact that 40 per cent of the world's populations now live in cities. By 2000 AD it is expected to reach 80 per cent.

Again and again it is this question of population. There are too many people on earth. They use available resources too fast and lay waste the world too ruthlessly. (In the U.S. alone, 7 million cars, 20 million tons of paper, 48 million cans and 26 million bottles are junked each year. With only 5.7 per cent of the world's population, the U.S. consumes 40 per cent of the world's production of natural resources. If the rest of the world used steel, for example, at the same rate as the U.S., all known reserves of iron ore would be consumed *in one week.)* Yet it is as a generator of stress that overpopulation is an even greater threat, because it magnifies all the problems created by an urban environment.

Much of Man's ingenuity and energy have been directed towards raising the level of life and eliminating death. Famine, pestilence and even wars have been curtailed or eradicated. More people live longer lives, producing more children who live longer. Yet quality and quantity are incompatible; while the supply of people may be endless, the necessities of life are not. The only way to break the circle is to decrease the birth rate, since increasing the death rate is not acceptable.

Family planning programmes have already been initiated by a number of Governments. India is running a birth control campaign promoting both contraceptive and sterilization techniques. The trouble is that contraceptive techniques which work well among the more educated are far less successful with the uneducated or the peasant. To produce as many children as possible, above all sons, is the overriding object of life among primitive people throughout the world. The pill, coil and diaphragm are strange, difficult or uncomfortable. A high death rate has always placed a premium on having as many children as possible in the hope that some will survive. Concepts like that die hard. Primitive agricultural communities need as many children as possible to provide labour on the farms and to look after the parents in old-age. Birth control programmes in developing countries that are trying to introduce the idea of the two-child family meet deep-running resistance. Government incentives are needed to encourage birth control. In India transistor radios reward those men that are sterilized, and this campaign is having some success. It is proving more popular than contraception, since all that is required is a simple operation. But most men only become sterilized when they have already fathered several children

Religious problems make the introduction of contraceptive techniques into Roman Catholic countries difficult and abortion as a method of birth control is hindered by Catholic and Moslem attitudes. Abortion has proved highly successful in Japan as a means of limiting the population, but the technique is, of course, feasible only where there is an advanced medical service available. The combination of sterilization and improved contraceptive techniques provides the best hope for the future.

Improving surroundings and limiting population to enjoy them — both massive, world-wide tasks — must be carried out immediately and simultaneously if the human race is not to render life unlivable. One without the other will fail. And if we fail, suicide by social stress or starvation is waiting round the corner.

IX

TOO FEW WILD ANIMALS

*And I think in this empty world there was room
for me and a mountain lion.
And I think in the world beyond, how easily we
might spare a million or two of humans
And never miss them.
Yet what a gap in the world, the missing white
frost-faced of that slim yellow mountain lion.*

D.H. Lawrence,
The Mountain Lion

Man is part of nature, he is not above it nor divorced from it but dependent on it for his well-being and, in the final analysis, his life. The seemingly all-embracing, self-contained, man-made environment of city dwelling tends to obscure this truth. Yet the very stresses of this artificially maintained life make it all the more important that Man should go out into the country for rest and renewal. The mass exodus from towns on summer weekends indicates an awareness of this, even if all he sees of the country is a few dusty blades of grass in a lay-by off the main road. He has changed from nomadic hunter-gatherer to settled farmer, and from settled farmer to urban Man. The communities in which he lives have grown ever larger and more complicated and, in the past 200 years, more and

POLLUTION: THE WORLD CRISIS

more divorced from nature. But 200 years are not enough to change the genetic inheritance of thousands of years. Man has been a part of nature for too long to be able to survive without it. He is dependent not only physically but also psychologically, and if he is to survive he must take steps to safeguard the natural resources — the land, the sea and the air — of the world around him.

So far 150 different species of birds and mammals have been exterminated, and 835 are in danger of extinction at the present moment. A sorry record. The phrase 'dead as a Dodo' means just that; the Dodo is dead — so are the Quagga zebra, the blue antelope, the Labrador duck, the great auk and many other species. The blue whale, the largest living mammal, indeed the largest creature ever to have lived on earth, yet a totally inoffensive animal, has been reduced to such an extent that the population may never recover. These animals could still be delighting the hearts and feeding the bodies of Man today, if past generations had refrained from wholesale slaughter. And what will our descendents think of us? The 'Daddy, what did *you* do in the War?' after the World Wars will change to 'Granddad, what did you permit in the Peace?' Future generations may curse us not only on aesthetic and cultural but also on economic grounds. Intelligent culling of surplus animals of any species allows Man to tap at will a truly inexhaustible food supply. Gross overhunting for a short period — and one hundred years is a short period — can wipe out a species for all eternity.

TOO FEW WILD ANIMALS

Setting aside all moral, ethical and psychological reasons, it does not make sense either economically or ecologically to exterminate whole species. Intelligent conservation is a policy that is soundly based on the knowledge of the long-term benefits to be gained by maintaining our reservoir of wildlife. In Russia 300,000 Saiga antelope can now be safely cropped. This species was once nearly extinct. By 1920 barely 1,000 had survived and the people of Kazakhstan who depended on the herds were suffering. Then the Russians passed a law to protect the antelope and established a breeding and rearing station. The policy has been so successful that they now number some three million head and the otherwise barren lands of the tundra have been turned into an economic asset. The protection measures taken by the Soviet Government have restored elk, sable and beaver populations which were all almost extinct at the beginning of the century. Fur production in the USSR now amounts to more than 50 million pelts a year. Conservation is a profitable business, not a matter of sentimentality.

The same pattern has been repeated in Africa. The creation of game reserves has not only helped preserve many unique species for posterity but has also proved a fruitful source of income. The tourist industry brings much needed foreign currency into these underdeveloped countries and a valuable reserve of protein is being maintained. It has been found that wild animals are a better protein source than the domesticated animals that have sometimes supplanted them on their home grounds. The wild animals have adapted to African conditions over the centuries and make the most efficient use of the land available. The savannahs of the Congo and Uganda can support a total weight of 40 tons of wild game per square kilometre, whereas only five and a half tons per square kilometre of domesticated animals can be maintained in the same area. For these perennially hungry countries this is a very valuable food source. The Zambian Government, in a country chronically short of protein, are finding their game-cropping programme of great economic value, as well as essential for the continued life of herds of elephants, hippopotami and buffaloes in the Luangwa valley. In its 3,500 square miles there are 22,000 elephants, 15,000 hippos and 12,000 buffaloes. It has been calculated that 6,000 elephants must be shot to reduce the population to a level that the valley can support. In 1968 the game-cropping programme produced £90,000. The conservation policy of game reserves has been, in a sense, too successful. The parks are small and limited compared with the original area the animals ranged freely over. Food supplies are consequently limited but, in their protected state, the populations expand enormously. In the Serengeti National Park in Tanzania there are now approximately 2,000 elephants: before it was set up there were almost none at all.

Conservation, therefore, does not mean shutting the animals up in game reserves and allowing them to increase indefinitely. This policy of total protection — pursued in the early days of game reserves — proved unworkable. According to a 1969 report by the consultant ecologist, C.A.R. Savory, in *Oryx*, the Fauna Preservation Society's magazine, the Rhodesian national parks face ecological ruin because of overgrazing. The animals have increased to such an extent that the pasture available is not sufficient to feed them and they starve. Over 30,000 head of game in the Tuli Circle National Land alone were dead of starvation.

Savory outlined six stages in the destruction of the habitat. Stage O: a stable balanced population of a wide variety of animals with hunting forbidden. Stage 1: mammals previously killed by man increase rapidly in numbers. This means that the perennial grasses are overgrazed and, as a consequence, their wide-spread root systems, essential to withstand dry spells, decrease. This decrease in competition allows the annual grasses, weeds and shrub seedlings to increase. Stage 2: areas of bare ground and shrubland grow. This is ideal for eland and impala but roan, tsessebe, rhino and reedbuck, with their natural habitat shrinking, decrease. The exposed soil is subject to erosion and rivers and pools carrying washed-away earth tend to silt up. The water supply for animals is reduced. Stage 3: a serious situation develops. The lack of grass drives the animals to eat mature trees. Stage 4: shrubs die off, because they have been overeaten and also because of the low water table. This exposes more land still. Vegetation is now annual and it becomes increasingly dependent on the rainfall, because annual grasses have shallow roots and cannot survive long dry spells. By the end of the dry season nearly all the ground cover has gone. Having nothing to eat, rhino, roan, tsessebe, bushbuck and reedbuck have disappeared. Stage 5: the trees are destroyed and desert conditions begin to encroach.

These effects are being seen in other national parks. In South Africa's Kruger National Park, elephant, impala, wildebeest, zebra and buffalo are all being cropped because they are too numerous for the available water supplies. This not only keeps the populations at manageable levels but the programme also yields valuable scientific information and economic benefits in the form of meat, leather and ivory. Elephants in the Tsavo National Park in Kenya are so numerous that they are destroying their habitat. They tear down the trees in their efforts to get at the leaves, this allows the grasses to grow in their place. Following the pattern established by Mr Savory, food supplies dwindle and the elephants will eventually starve if not cropped. An accidental benefit of this process has been a marked increase in agricultural productivity. Grass fires clear the fallen timber and the bush is slowly being turned into parkland. As grass replaces bush, the tsetse fly, a cattle scourge, is deprived of its natural habitat and dies out. In this way it has been virtually eradicated in the Tsavo Park. In the past, highly desirable fly-free pastures which could support hugh herds of cattle have been created in this way.

TOO FEW WILD ANIMALS

South Africa has more than 200 public game and nature reserves covering 51,766 square miles. She has been outstandingly successful in the field of wildlife conservation and was the first African country to show any interest in establishing national parks. In 1969 they succeeded in re-establishing the white rhinoceros in the Umfolozi Game Reserve in Natal. For some years this had been an endangered species but in that year it was declared out of danger. The recovery has been so successful that 40 white rhino were transferred to the Kruger National Park and others to the Gorongosa and Maputo Game Reserves in Mozambique.

The efforts of the conservation-minded African states to build up and manage their game reserves are plagued by the activities of tribal and organised poachers. The Albert National Park in the Congo, once the pride of the Belgians — where, since independence, effective protection has ceased — has been almost completely destroyed. In Zambia the black lechwe antelope — one of the rarest specimens of game in the world — is faced with extinction, as is the crocodile in Kenya which is hunted for its skin. The 'spotted cats' — leopards, jaguars, ocelots, cheetahs, etc. — have also been reduced to a fraction of their numbers because of the demand for their fur. The profits to be made by poaching have meant that an immense illegal trade in these furs has sprung up in Africa, Asia and the Americas. Despite the efforts of the governments of both the supplying and receiving countries to control this trade it will continue while women are prepared to wear the skins of these animals. Women who wear real fur coats made from the skins of endangered animals when modern artificial furs cannot be distinguished from the real thing, should be objects of scorn rather than envy. While the demand for real furs continues, men will supply it, and feminine snobbishness and vanity may exterminate many valuable species. In 1969 a tiger skin coat made for Gina Lollobrigida used 10 skins — one-sixtieth of the total world population. As Peter Scott said, 'They look better on the animal.'

The maintenance of the animal population of Africa is not the concern of the African nations only. President Kenyatta of Kenya pointed out at the Arusha Conference of 1961 that if the wild life of Africa is to be preserved as a world heritage the world must be prepared to pay for it. A fair comment. One response was the World Wildlife Fund (WWF), set up in 1961 and concerned with fund-raising, publicity and co-ordinating all conservation efforts on an international scale. The headquarters are at Morges, Switzerland, and branches of WWF exist in all Western and many African and Asian countries. The WWF provides monetary support and scientific advice for the conservation efforts of many governments. It was mainly due to WWF efforts that Ethiopia's first national parks were established in 1969. Its campaign to save the 'spotted cats' has already had some success on a governmental level. In June 1969 they appealed to the European and North American

Governments to control the import of spotted skins. In July Canada forbade the import of the skins of nine species of cats including the leopard. In December President Nixon signed a law forbidding the import into the U.S. of any animals, or parts of them, that were in danger of extinction. The International Union for the Conservation of Nature and Natural Resources (IUCN) lists some 835 species of birds and mammals that are menaced, partly due to the destruction of their natural habitats and partly to direct action by Man. If all governments followed the American example, a major step towards preserving these species would be taken.

Where animals have not been hunted for some desirable product — beautiful skins, feathers or mythical aphrodisiac qualities — they have been exterminated for sport or apparent convenience: usually at the cost of ecological imbalance. In South Africa the hippopotamus was deemed a useless animal that cluttered up rivers: hippos were shot on sight. As a result bilharzia (also known as schistosomiasis), a debilitating disease of the tropics and sub-tropics and today the commonest disease in the world, has spread. Subsequent research found that hippos helped keep the rivers clear by stirring up the silt, which maintains deep, fast-flowing channels. When the hippos were eliminated, the rivers quickly silted up and the adjacent lands were periodically flooded. These conditions favoured the growth of the water snail that carries bilharzia.

As Man learns more about ecology, the vital

importance of one animal or plant is sustaining a healthy population of all animal and plant life is becoming clearer. The interaction of all forms of life, whether on water or in air, contributes to a healthy environment. Remove one factor in this ecological balance and the whole process gets out of gear.

The complexity of the problem is clearly shown by the state of the reindeer population on the west coast of North America. Their numbers were originally kept down by their natural predator, the wolf. When this was eliminated by intensive hunting, the reindeer, with nothing to check an increase in population, multiplied prodigiously. The consequent destruction through overgrazing of their habitat has brought about starvation conditions and a severe decline in numbers. The same thing may well happen in the Arctic to the caribou where, until recently, bounties were being given for all wolves killed. At the Ninth International Conference of Game Biologists held in Moscow in September 1969, it was reported that a study of deer killed by wolves in Siberia's Taimyr Peninsula had shown that more than two-thirds of the deer were suffering from diseases of various kinds. Boris Bogdanov, head of the Soviet Nature Conservation Board, stressed that elimination of predators had harmful consequences. Disturbance of the numerical balance between wolves and deer causes an appreciable increase of disease among the deer and can even lead to epidemics. In the Crimea where the wolves had been eliminated, the deer were becoming slower, had less endurance and their antlers were smaller. For it is the predators that kill the weak, the sick and the old in their balanced environment and thus help to maintain healthy populations.

The otter is another case in point. It was virtually eliminated in Europe because of the damage it was supposed to cause fish populations. But fishermen found that once the otter had been removed, the fish population after a temporary increase declined. It was then found that the otter preyed mainly on old and diseased fish thus maintaining a healthy population. Once the predator was removed, disease spread and the overcrowded population was weakened. A special campaign for the protection of the otter has been launched by the British branch of WWF. Another reason for this gross overhunting all over the world, is the value of their skins. The IUCN'S Legislation Commission is currently working on a world-wide convention on the import and export of these animals. Here again widespread adoption of measures like President Nixon's to ban imports would go far to preserve and increase the world's otter population.

When effective protection measures have been introduced in time, wildlife populations have recovered. The sea-otter, which was nearly extinct at the beginning of the century, has now risen to about 36,500 and is almost completely re-established in its former territories on the Pacific seaboard of North America, from the Alaskan mountains down to the Californian coast. The Alaskan fur seal from the Pribiloff Islands in the northern Pacific, almost

POLLUTION: THE WORLD CRISIS

extinct by 1911, now yields 60,000 high-quality skins a year, and provides a valuable source of income to these islands.

The plight of seals in general has aroused world-wide concern since photographs of baby seals, being clubbed to a bloody death in their thousands, began to appear recently in newspapers and magazines: although new hunting regulations announced by the Canadians in 1969 were aimed at eliminating some of the more obnoxious aspects of this annual slaughter. The Canadian Government delayed the start of the hunting season by two weeks to allow the helpless whitecoats to reach the 'beater' stage where they are old enough to take to the water and have some chance of escape. Nevertheless, first reports in 1970 from the Gulf of St. Lawrence showed that this slaughter was as 'brutal, bloody and cruel as before'.

The hunting of seals is not, *per se*, harmful to the population; indeed to ban hunting altogether would probably lead to the same overpopulation problems witnessed in other wildlife species. It is where this hunting is completely uncontrolled and left only to the greed of the hunters and the appetite of the consumer that permanent damage to the species as a whole happens. In the Gulf the Canadian Government, under whose jurisdiction it falls, has set the maximum sustainable yield of baby harp seals at 60,000: that is the amount it is safe to kill every year and still maintain population levels. On the Labrador Front, however, off the Labrador coast outside

Canadian territorial waters, there are no such restraints. Combined Canadian and Norwegian hunting is at present killing three times the maximum sustainable yield of 40,000. The result has been a decline over the whole area from an estimated three million in 1951 to two million in 1969. Hence the very existence of these herds is in danger. Here again is another case where cropping the excess population ensures the maintenance of a healthy reserve of seals for future generations while over-hunting, though yielding a short-term financial advantage, will lead in the end to the extinction of the species.

Turtle soup is already an expensive luxury and soon it could become only a memory. All the seven species of marine turtle have reached a critical stage. They have been hunted both in their breeding and their feeding grounds because every part of the turtle is useful: eggs, meat, shell, leather, cartilage, and young shells for curios. Despite official protection in many countries, turtles in the Caribbean, South America, Costa Rica, Malaya and Sarawak are all running captive rearing programmes in an effort to build up the declining populations, but a complete and effective ban on all hunting is needed to help restore the turtle population to a safe level.

That the maintenance of reserves of wildlife all over the world is of great economic value is beyond doubt. But there are also less tangible benefits to be gained by preserving the fauna of the world in their natural habitats. Man once fitted neatly into his own ecological niche; he has broken out of this to become the world's dominant species. Genetic adaptation to the urban environment over thousands of years may in the end produce a species not very closely resembling the human race of today. But for the time being Man is still part of nature and still needs to keep in touch with it, if only symbolically. 150 million people visit the national parks of the U.S. each year. Most of these visitors live in overcrowded cities where fresh air is unknown, water is chlorinated, food processed, where Man the food-gatherer gathers only money, and the seasons are marked only by the switch from central heating to air-conditioning and from football to baseball on television. Men seek escape from their artificial life, and the national parks help meet this need. The U.S., the greatest exponent of urban living, was the first country to establish national parks. In 1872 the 3,200 square miles of the Yellowstone National Park were set up; since then the National Park Service has established 34 more, covering 14,453,000 acres.

That conservation interests are becoming more powerful was shown by the case of the Everglades National Park in the southern tip of Florida. This unique ecological complex is made up of many different habitats: coastal mangrove swamps, huge sawgrass savannahs, and inland pine and hardwood plantations. In the late 1960s the need for a new international airport became acute. The site chosen by the Dade County Port Authority was six miles north of the Everglades' borders. Now, although the

TOO FEW WILD ANIMALS

Park covers some 1.4 million acres, it is not ecologically independent: it relies on water from beyond its limits. As much as 38 per cent of the Park's water is supplied by the Big Cypress Swamp, just north of the Glades which was being bought up by the Port Authority as part of the scheme. Already vital water was being diverted to satisfy the increasing demands of Florida's farms and growing cities. The destruction of the Swamp would have meant a crippling blow to the Everglades — the further dehydration of the area and an assault on all the animals and plants living there. There were 22 endangered species to whom the Park was one of the last refuges. They included the bald-headed eagle, osprey, Florida panther, alligator, and the manatee — that huge, ugly but gentle sea cow which inspired tales of the mermaid. And on top of all this, the noise and air pollution. The battle-lines were clearly drawn between the conservationists (backed by the Secretary of the Interior) and the powerful State and commercial forces demanding the airport. If the threat had come as little as two years earlier, before the popular awakening to the importance of conversation, it seems unlikely that the White House would have intervened.

National parks and game reserves in all countries are also valuable from the scientific viewpoint. The International Biological Programme (IBP) started operating in 1967, one of its major objectives being the understanding of the complex inter-relationships in 'virgin' ecosystems, i.e. ecological areas that have not been grossly disturbed by Man. The parks and reserves can be used for this purpose. Detailed analyses of many different ecosystems are being made. As suggested by IBP's sub-title, 'Biological Productivity and Human Welfare', once the functioning of the world about us is better understood, the knowledge can be used to increase food production. In the past, ignorance of just these sort of ecosystems has lead to such fiascos as the East African ground-nut scheme and the creation of the dust bowls of the U.S. Planetary management must be the aim of Man but for this much more knowledge is needed. The inter-relationships in biological communities have evolved over millions of years until a perfect balance has been achieved. Man has to transform the environment to increase productivity but only with understanding can this be done non-destructively. As Professor Jean Dorst, writing in the *Unesco Courier* of January 1969 said, '[Man] has realized that nature conservation is not solely the domain of a few biologists preoccupied with protecting some rare species of animals, but is in fact the far greater problem of the rational use of mankind's natural capital and living space. A well-planned management policy can still safeguard this capital while substantially increasing the income from it to meet Man's legitimate needs.'

X

ENGINEERING HAVOC

Man has always modified his environment to suit himself. Where his changes have not been too violent there has been little long-term harm. Indeed, in many cases the results have been beneficial to both Man, animals and the land.

He first began to cultivate the soil about 10,000 years ago in the Neolithic Age. To start with, the space he occupied was negligible. He cleared small patches of forest on hill tops because the soil there tends to be lighter and easier to work and his equipment was not strong enough for anything else. When the land was exhausted in five to ten years, he moved on; forest grew back and the soil was replenished. With technological advances — the plough, the yoke and irrigation equipment, for example — Man could exploit the valley bottom soils, kept fertile by the periodic inundation of silt-rich flood water. The need for constant communal effort and co-operation to control these floods and work the land was one of the catalysts for the early civilizations of the Nile, Euphrates and Indus valleys. Small villages evolved into cities and the cycle of exploiting the world around began in earnest. But even where his activities damaged the environment, as long as the population remained low and therefore the area affected not too large, it would in time be repaired by nature.

POLLUTION: THE WORLD CRISIS

But since 1600 the human population of this world has grown remorselessly fast. From some 500 million four centuries ago, it has swollen seven times to 3500 million today. By 2000 AD it is expected to reach 7000 million. With an increase in population as vast as this, all available resources will be stretched to the limit to provide food, water, minerals, fuel and other necessities for Man. There will be no room for wastage. Even to satisfy the basic need for housing for this number of people will be a serious drain on available land. The great majority of people will have to accept living in more and more tightly packed towns and cities.

With an urgent situation like this it is essential to avoid, and if possible to make good, the mistakes of the past. The damage inflicted in the last two centuries in the frantic search for food and mineral resources has been immense. The combination of ignorance and short-sightedness has caused the loss of an estimated 1250 million acres of arable lands through erosion and salinization alone. The land has been torn up in efforts to extract its mineral wealth and enormous devastated areas produced. According to John Barr, author of *Derelict Britain* some 250,000 acres in Britain are derelict and this is increasing at the rate of ten acres a day. The deterioration of the environment has been caused by a combination of three factors: increased population, increased urbanisation and modern technology. Technology may have developed new and efficient methods of exploiting the earth's vegetable, animal and mineral natural resources, but all too often Man has used them only to gain in the immediate short term. The maximum possible profit has been wrung from the lands and seas with no thought for the future. The ecological consequences of nearly every project Man has undertaken have been ignored. The world has been there to take and use in whatever way was most profitable. The difference between the situation today and even 40 years ago is that one more factor has beeen added to the profit and loss account and that is the escalation of pollution.

A case in point is the vast new irrigation schemes in Africa, among them the Aswan High Dam, Niger-Kaduna, the Kariba Dam and the Gouda, designed to produce food for an ever-multiplying number of hungry mouths. They were intensively researched from the economic and engineering angles but not from the ecological. Take the Aswan High Dam. It was intended to provide 2,100 megawatts of electrical power to encourage the setting up of light and heavy industry and a 25 per cent increase in cultivated land in Egypt through desert reclamation by irrigation. The two vital stimuli to an industrial revolution combined in one project. There is, however, one basic, catastrophic flaw in the whole concept: the dam holds back most of the rich silt on which the farm land downstream depends, not only for its yearly fertilization but also to offset natural erosion. Nor is this all, for the dam is also having deleterious effects on the fisheries of the Nile delta.

Before the building of the dam the nutrients washed down the river kept these fisheries highly productive. As the nutrient content declined, so did productivity. Only three years after the dam started operating in 1967 the value of the sardine industry has dropped by $7 million (the catch has declined from 18,000 tons in 1965 to 500 tons in 1968) and shrimps are going the same way. Further, the dam has encouraged the spread of bilharzia, a debilitating disease endemic in Egypt. Present when the land was irrigated by seasonal flooding, the disease is flourishing now that the land is perennially flooded. In the delta nearly 100 per cent of the population is suffering from it. The disease is carried by freshwater snails and passed to Man, where the eggs, larvae and worms live and breed in the bladder and intestines, causing diarrhoea, anaemia, fever and many other unpleasant and dangerous side-effects. Hence reservoirs have silted up, delta lands been lost, salinization increased and water-borne diseases such as bilharzia have spread. All this apart from the suffering caused to people who have been displaced from their ancestral lands by the rising water.

The building of the Kariba Dam on the Zambesi, which started operating in 1959, has had some unfortunate results. Studies carried out there by Parsatam Hira, then of the University of Zambia, showed that here too bilharzia is on the increase. Prior to the creation of the dam it was not prevalent in the region, but the 2,050 square miles of fresh water formed by the lake has created ideal conditions for the snail, particularly in the inlets and shoreline where the flourishing water weed *Salvinia auriculata* keeps the water calm. Destruction of the snails' habitat should help control the disease.

Irrigation can also result in salinization and, curiously, render vast areas of land useless. By definition irrigation takes place in dry areas and, because of the high evaporation rate, these have high concentrations of salts in the soil. As the land is flooded these salts dissolve in the water. When irrigation is finished and the water recedes they tend to be precipitated *on the surface.* Such an artificially high and locally concentrated level of salts makes plant life impossible. A further risk is that when the water is used to irrigate fertile and salt-free land, it will contaminate it and cut its productivity also. The British carried out irrigation works in the Indus valley, West Pakistan, at the beginning of the century. This did not result in the expected rise in food production, for seepage from the unlined canals caused the level of the underground water table to rise. This not only carried salts to the surface of the soil but also waterlogged the land. In the 1950's 100,000 acres of land a year were being lost because of this. The situation is now being remedied. With international aid from the Colombo Plan, the West Pakistan Water and Power Development Authority is virtually rinsing the land by pumping fresh water from 100-foot-deep tube wells to clean the soil and remove the saline water.

Man must learn from these past failures. Much as

arid lands need water, irrigation schemes may merely produce one fertile area at the expense of another or, worse, merely destroy one fertile land without creating another.

The largest, perhaps the most ambitious, engineering and irrigation scheme the world has ever seen, affecting half a continent, is planned by the Russians. The scheme, announced in February 1970, could have world-wide repercussions. The plan is to divert the waters of the Ob, Pechora and Yenesei Rivers, which at present flow north into the Arctic Ocean, into Central Asia where they will be used to irrigate some 85 million acres of arable land. At the same time the canals which carry the diverted water should help drain vast areas of swamp land thus producing another 150 million acres of arable land. This should help slow down the rate at which the levels of Lake Balkash and the Aral and Caspian Seas are dropping. At present their waters are being used for irrigation purposes and the Aral Sea is in danger of drying up altogether. The Pechora will be dammed so that its level rises by 130 feet, then a 70-mile canal will carry the water to the Kama and thence to the Volga. This project should take 15 - 20 years. The Ob will be turned round by building a huge dam near Tobolsk, at the confluence of the Irtysh and Tobol. The Tobol will be forced to flow backwards and a network of canals will be built to carry 10 to 12 cubic miles of Siberian water down to the Aral Sea basin each year. The three main canals will be from 400 to 550 miles long. This part of the scheme should take 15 years to complete.

On the face of it the advantages are obvious: a total of 255 million acres of arable land opened up and at the same time the problem of draining west Siberia neatly solved. Immense deposits of oil and gas have been discovered there but they are difficult to extract because the land is so marshy. But the implications for the world are staggering. Because less fresh water will reach the Arctic Ocean less ice will form, hence the sea will lose more of its heat, the North Pole will warm up and the ice cap may shrink. As a result the climatic zones of the northern hemisphere may move north by as much as two to three hundred miles. This could mean that all borderline arid or semi-arid land (and Russia has great tracts of it) would be turned into desert and Sahara-like conditions could spread from North Africa into Spain, southern Italy and Greece, although the Mediterranean Sea should moderate this effect. It would also mean that Britain could benefit from a warmer climate and hotter summers. There has been no general agreement among meterologists on the climatic results. When the consequences of such a major engineering operation are impossible to predict it seems unwise to meddle with the present balance. However, the Russians are going ahead with the scheme and in some 50 years or so we should know.

The proposed new Panama canal between the

eastern Pacific and the Caribbean could also have some undesirable ecological consequences. The plan under consideration is to excavate a sea-level canal across the Panama isthmus using atomic explosions. The results of allowing an uncontrolled biological exchange between the different marine species of the two oceans are not clear. However, the introduction of foreign species into 'virgin' environments has often had disastrous consequences. The natural checks and balances that control the populations in their normal habitats do not exist. One possibility is that a relative of the crown of thorns starfish *(Acanthaster planci)* that is currently devastating the coral reefs of the western Pacific will spread to the Caribbean. This species *(A. ellisi)* of the eastern Pacific is at present limited in numbers, but a sea-level canal would allow it free access to the rich coral reefs of the Caribbean and the epidemic could spread. Another undesirable species — the poisonous sea snake — could invade the Caribbean via the canal, endangering both marine and human life. Should the canal be built, biological barriers must be incorporated to prevent free migration of the different marine species.

On a smaller scale Man has already seen the devastation caused by a misapplied technology. The scars of the industrial revolution are still with us today to be seen in the north of England, the mining districts of Germany, the Borinage of Belgium, the hills of Kentucky and many other areas of the world. The misuse of farming techniques have rendered vast areas of the world barren (the dust bowls of the

ENGINEERING HAVOC

American Mid-West were caused by ploughing the prairie too often) and feckless exploitation of water resources has caused still more damage. None of this was necessary. Technology need not result in pollution of the waters, destruction of the land and pillaging of the seas. Neither increasing population nor expanding technology has to destroy the land which both need to survive. It is the rapist attitude of industry and Man that has caused and is causing the damage. Each generation has behaved as if it were the last one. Today's generation wants tiger-skin coats at the cost of eliminating all tigers. Whale hunting is a profitable industry — harpoon them all: the stocks may soon be exhausted but for a few years more some men will have made a lot of money. Industries operated with no form of pollution control mean maximum profit for shareholders and owners. We cannot afford this attitude of mind any longer. Sound ecological principles must be applied to the future exploitation of earth's resources if Man is to survive. We usually have the knowledge and always the power to modify the environment to suit ourselves without at the same time rendering it uninhabitable both for Man and other animals. We must learn to apply the power of technology beneficially.

XI

HOW THE WORLD MUST ACT

'The great question of the '70s is: Shall we surrender to our surroundings or shall we make our peace with nature and begin to make reparations for the damage we have done to our air, to our land and to our water?' These words of President Nixon in his State of the Union message in February 1970 are applicable to every country in the world. Pollution is now a world crisis — all countries are now paying the price for the destruction of land and water resources and the fouling of the air. The probable consequences of allowing the invisible pollutants — carbon dioxide, DDT, etc. — to increase have at last been evaluated. For the first time the 'man-in-the-street' is expressing concern about the quality of his environment, and many governments are now at least paying lip service to the ideas of conservation and pollution control. In the U.S. pollution is replacing Vietnam as the burning issue on the college campuses. (April 22nd 1970 was designated as 'Earth Day' — a day on which campuses across the country organised lectures, symposia, demonstrations and marches in an attempt to bring home the facts of the environmental crisis to the whole country.)

Concern about pollution cuts across all barriers of race, age and occupation. In the past few years a handful of militant ecologists have done much to bring the issue before the general public. Prominent among them have been Barry Commoner of Washington University in St. Louis, René Dubos (Rockefeller University), Paul Ehrlich (Stanford), Frank Fraser Darling of Great Britain and politicians Senators Edmund Muskie and Gaylord Nelson — all of whom have been 'prophets crying in the wilderness' for many years. Pollution has now become a major political issue in the U.S. Demonstrations against industries with bad pollution records have multiplied and the Santa Barbara oil disaster raised a storm of protest across the country. The growing volume of protests about pollution of all kinds led President Nixon to promise, in his State of the Union message to Congress, that he would make the fight for a better environment the major concern of the 1970s.

He outlined new standards for air and water control, new methods for dealing with solid wastes, and recommended an increase in the number of national parks and municipal recreation facilities. These plans will cost money. The Federal Government will spend $4,000 million between 1971 and 1975 on the improvement of existing sewage plants, construction of new ones and control of industrial and agricultural wastes. State and local governments will be expected to find $6,000 million for the programme. Nationwide uniform water purity standards must be established and penalties for pollutors increased. A fine of $10,000 a day for pollutors of streams and lakes is proposed. A similar penalty is suggested for those who transgress the air quality standards. Again, national air cleanliness standards are proposed and tougher car exhaust controls are being introduced. Mr Nixon has expressed the hope that a pollution-free vehicle can be produced by 1975. He has also proposed that the initial purchase price of a car should include the cost of getting rid of it: one of the U.S.'s worst eyesores is the motor junk-yard full of rusting, battered hulks, piled high on top of each other, that is a depressing feature of nearly every city, town and village. Research into producing both readily disposable and degradable products − especially containers − and re-usable articles will be stimulated. He also intends to re-organize the Federal agencies more efficiently and to mobilize municipal and industrial groups to co-operate with the newly established Council of Environmental Quality.

In Britain, too, pollution has become a cause for major public concern, although in the past vigilance on pollution has been greater than in the U.S. In October 1969 Mr Anthony Crosland, as a Cabinet Minister, was made responsible for dealing with environmental pollution as part of his job as Secretary for Regional Planning and Local Government, and the year ended with Mr Wilson setting up a Royal Commission on Environmental Pollution. However, the new Conservative Government only made a Parliamentary Secretary to the Ministry of Housing and Local Government responsible for the environment and it only merited one paragraph in the Queen's Speech. In February 1970 both the House of Commons and the House of Lords debated the subject of air and water pollution in Great Britain. Lord Kennet, Parliamentary Secretary to the Ministry of Housing and Local Government, said that the present penalties for pollution were derisory and should be brought to more realistic levels. Lord Molson said that the money being spent on purifying sewage, at present about £100 million a year, ought to be trebled to about £300 million. The existing facilities were inadequate, 60 per cent of present sewage treatment plants sub-standard, and the present laws not being enforced properly. The Clean Air Acts that had effected such a marked improvement in the quality of London's air still were not being applied in some areas. The problem of Britain's air pollution affects people living not only there but also in Europe — particularly Scandinavia.

The international aspect of pollution was one of the main points stressed at the European Conservation Conference held at Strasbourg in February 1970. This conference marked the start of the International Conservation Year and was convened by the Council of Europe to discuss 'the management of the environment in tomorrow's Europe'. The draft declaration approved by the conference included the proposal that the European Convention on Human Rights should be extended to guarantee 'the right of every individual to enjoy a

POLLUTION: THE WORLD CRISIS

healthy and unspoiled environment. This should cover the rights to breathe air and drink water reasonably free from pollution, the right to freedom from undue noise and other nuisances, and to reasonable access to coast and countryside.' Among the important points that were agreed by the conference were: 'Policies should be strengthened or introduced to control pollution of air, water and soil, and internationally agreed standards for those purposes should be devised as soon as possible', and 'Rational use and management of the environment must have a high priority in national government policy ...'

Many impeccably sane recommendations were made and the phrases have a reassuring ring about them. But, as the Duke of Edinburgh pointed out, 'All the impassioned speeches will be so much effluent under the bridge unless it is followed by drastic political action.' Such action is essential. The conference had no power; it could only recommend guidelines. It is up to the governments of the 27 countries represented at the conference to take the necessary action.

International standards are needed for industry, for pesticide control, for noise and pollution by aircraft. All these are international problems — no one country can pollute the atmosphere without affecting its neighbours. Many rivers, such as the Rhine, flow through several countries, sweeping pollutants through them and out to sea. A supranational European body is needed, which would have power both to lay down pollution standards and to enforce them.

Pollution is a problem, then, which is not contained by conventional political boundaries — co-operation between the communist and capitalist countries is essential: and this co-operation is lacking. For example, East and West must curb pollution in the Baltic Sea; but, though both Russia and Sweden are agreed on the necessity for joint measures to combat the pollution, no active steps have yet been taken. Sweden convened a conference in September 1969 which the USSR, Poland, Finland, East and West Germany attended, but no useful results were achieved. They unanimously agreed that something should be done — but nothing was. The Western countries wished to deal with the problem at an administrative level, while the Eastern countries insisted on a ministerial approach which would imply recognition of East Germany. Thus even when all countries are agreed in principle, effective action to save the environment can be blocked by political differences. Yet it is surely worth surrendering some degree of national sovereignty to ensure the survival of an environment fit to live in.

This intensely selfish approach has delayed many anti-pollution measures. Governments have been loath to introduce anti-pollution legislation unilaterally because this would penalize their industries in the world market. Where adequate laws exist, industry has been dilatory in applying them because installing proper anti-pollution machinery

POLLUTION: THE WORLD CRISIS

increases their costs, forces up their prices, makes their goods uncompetitive and drives them to the wall. Firms which do not spend money on disposing their effluents are at an obvious advantage over those that do. This is why international standards must be established and applied so that no firm gains this unfair advantage over rivals.

For a foretaste of the sort of political and financial problems which pollution control, both national and international, will come up against in the future, take North America. It exemplifies, in miniature, the world-wide difficulties. States which rigorously apply the existing water and air quality control regulations tend to lose profitable industries. Companies simply move, with the jobs they provide and the taxes they pay, to other States less concerned with environmental problems. Thus President Nixon has had to establish nation-wide Federal standards.

The Federal Government in Canada is for the first time taking the initiative in a campaign to clean up Canada's inland waters. In August 1969 the Provincial premiers agrees to co-operate with the Federal Government in a nationwide battle against pollution. It might seem easier to get the Provinces in one country to co-operate in the common battle against pollution than to get the different nations of Europe to work together. But this is by no means the case. The Provincial Governments are all jealously trying to maintain their authority. The Federal Government, therefore, will be forced to let the Provinces take their own measures to fight pollution but will try to co-ordinate the various activities. Legislation is to be introduced committing the Government to a national water quality programme. It seems strange that such an enormous country with vast natural resources and a limited population should have a bad water pollution problem, but 60 per cent of the population is squeezed into the 35,000 square miles along the U.S. border and sewage purification plants are primitive, where they exist at all. 90 per cent of Quebec's sewage is untreated, and the pulp and paper industry is a major polluter. One pulp mill can produce as much effluent as the untreated sewage of a fair-sized town.

Northern Canada is particularly susceptible to abuse. With the extreme cold there, the natural processes that cut down the harmful effects of pollution, for example the degradation of oil, are very slow. Because of this the balance of its ecology is fragile and easily disrupted. As we saw in Chapter IV the trial voyage of the supertanker *Manhattan,* ending as it did with large holes in the hull which would have spilt thousands of tons of crude oil if the ship had been loaded, caused great concern in Canada. The Canadian Government has insisted on a number of anti-pollution measures being observed for its second voyage via the North-West Passage in April 1970. The lessons of the south have been learnt, and the exploitation of northern Canada is going to be governed by much stricter rules. The Minister of Northern Development, Jean Chrétien, outlined a number of major conservation measures designed to protect the environment in October 1969.

Canada's pollution problems involve co-operation not only between the Provincial and Federal Governments but also between the Canadian and U.S. Governments. The pollution of the Great Lakes is an international problem. Both the U.S. and Canada are planning an International Field Year for the Great Lakes when Lake Ontario will be studied in detail. There is still considerable disagreement on the exact processes that have caused the eutrophication of the lakes and no general agreement on what remedial action should be taken. Curbing the present sources of pollution would be a useful first step, and in a number of individual cases this has already started, but integrated action is essential if any lasting improvement is to be seen.

In March 1970 U Thant, Secretary General of the U.N., warned that life on earth could be preserved only by joint international surveillance, consultation and action on environmental matters. He was speaking at a conference in New York which laid the foundations for the U.N. Conference on the Problems of the Human Environment to be held in Stockholm in 1972. Sweden was one of the prime movers behind this conference, which is designed to provide a focus for world-wide action to avoid a possible crisis that could endanger the well-being of mankind. Acceptable international standards must be proposed and suitable monitoring systems worked out.

That the problems are global rather than even international is readily seen. Pollution of the atmosphere — by carbon dioxide, dust and radioactivity — and of the oceans by oil, pesticides and other toxic chemicals — spreads fast and cannot be confined. As the American Chemical Society Report pointed out life on earth could end if a contaminent destroyed any one of at least half a dozen bacteria involved in the nitrogen cycle. Even pollution of inland lakes and rivers becomes a global problem when that pollution affects the productivity of the estuaries and so the future yield of fish from the sea. In addition, wildlife conservation should be dealt with on a global basis: for nature reserves and game parks are a common world heritage. It has become clear that Man lives in *one* biosphere and few actions can be taken that do not have global repercussions.

Is there any hope that the U.N. conference will achieve something more than the reiteration of the problems we are already aware of and a spate of platitudes on co-operation? Only if the world is prepared to pay the price of pollution control. There are few environmental problems which cannot be solved using current technological knowledge. In fact technological 'solutions' exist to most industrial pollution problems. But they are not applied, usually for economic or political reasons. Pollution control is admittedly expensive and most industries are unwilling voluntarily to spend the money. They must, therefore, be compelled to do so by effective legislation. Since this increased cost will inevitably be passed on to the consumer, it will result in higher prices for industrial goods. Again, purifying municipal sewage is an ever-increasing burden which the ratepayer and the taxpayer will have to bear. It is an essential social service which must be operated at peak efficiency.

Car pollution is an area where everyone can demonstrate the sincerity of his desire for a clean environment. Stringent exhaust control measures and lead-free petrol will result in less power and increased cost, neither of which will be over-popular with the motorist but both of which must be accepted. It is little use complaining bitterly about a deteriorating environment if we are not prepared to pay the price for cleaning it up.

Despite our greatly increased knowledge of the environment, there are still many areas where useful research remains to be done. If international standards are set up, improved monitoring systems must be created so that the regulations can be enforced. More than that, there must be a concerted effort, common to all Men and every nation, to destroy the mass of undegradable containers — plastics, tin cans, rusting cars etc — that defile the earth. In addition, biodegradable materials must be developed which are easily broken down, and any long-lasting materials must be capable of re-use. Given the limited supplies of raw materials on earth, methods should be developed for breaking down and recycling used articles. It will not be many years before cars will no longer be allowed to rust uselessly away.

HOW THE WORLD MUST ACT

The problems that face Man today are many and great, but they are not insuperable. We have the knowledge and the power to right many of the wrongs done to the environment in the past, and we can ensure that these mistakes are not repeated in the future. If we do not take action immediately to stop further pollution and clean up the mess already made, Man is doomed. He can die in a number of ways — radioactive poisoning, suffocation, starvation — because he has destroyed the lands and seas on which he lives. The stresses of overpopulation could lead to a complete breakdown of our present civilisation.

Failure to control its environment characterizes a civilisation on the decline. The horrific and sudden cataclysm of a nuclear war which will utterly destroy our world is a common nightmare. But we must realise that the slow chronic destruction of the environment is just as dangerous — perhaps more so; because of its slowness and because we all contribute, we all have our finger on the button — every day in every thing we do. We cannot hide from our responsibility by saying that our individual contribution is too small to matter. It is our own neglect that is fatal, not someone else's. Everyone must choose — for themselves.

GLOSSARY

bilharzia	see schistosomiasis
biodegradation	the breaking down of a compound into its component parts by biological processes
biological magnification	the progressive concentration of nutrients etc. up the food chain
biosphere	the surface layers of earth where life exists including air, land and water
carnivore	animal-eating animal
DDT	dichloro-diphenyl-trichloro-ethane
ecology	study of the habits of living organisms and their relation to each other and their environment
environment	everything in its surroundings that influences a living being
eutrophication	over-feeding of the water leading to premature ageing
food chain	the prey-predator relationships from small animals preying on smaller, to large animals preying on small, plankton — small fish — large fish — birds — larger birds — birds of prey
halocline	water level at which there is a sharp change of salinity, impeding the exchange of water below and above

GLOSSARY

herbivore	plant-eating animal
meteorological inversion	condition when the upper layers of air are heated trapping cool air beneath
photosynthesis	process whereby carbon dioxide and water are converted into glucose and oxygen
phytoplankton	microscopic marine plants and animals
schistosomiasis	water-borne disease carried by snails, also known as bilharzia

INDEX

Alaska	51
algae	12,17,26,159
American Chemical Society	23,26
antelope, black lechwe	131
Saiga	127
Ardrey, Robert *The Territorial Imperative*	119
Arctic Task Force	52
atmosphere	1,7,85
bacteria	21,22
Baltic, Sea	32,39,62,74,155
Barr, John, *Derelict Britain*	142
beaver	127
bilharzia	133,144
biological control	47,78
magnification	65,68
biosphere	1,7,9
bronchitis	96
Buckminster Fuller,R.	vi
calcium	65
carbon cycle	2,3
dioxide	2,7,8,85,87
monoxide	88-9,98-101
cars	87-9
electric	91
Carson, Rachel *Silent Spring*	70
cats, spotted	132-3
Caspian Sea	18
caviar	17-18
Cheshire, Dr Richard	7
chromatography	56
coho salmon	71-2
crocodile	131
crown of thorns	6,7,148
cytoplasmic incompatability	77

INDEX

Dam, Aswan High	146-7
Kariba	147
DDD	70,73
DDE	67,70
DDT	61-3,66-76,79,81
banned	81-2
deoxygenation	9,12
detergents	12
dieldrin	62,67,82,84
eagle, bald-headed	66
golden	66
elm bark beetle	68
estuaries	33,38,73
earthquakes	111,114
earthworms	68
ecology	7
ecosystems	5
elephants	128,130
eutrophication	12,17,22,26
falcon, peregrine	66
family planning	123
Federal Water Pollution Control Administration	15
fertilizers	12,22,25,27
fluorine	92
fluorosis	92
Food and Agriculture Organisation (FAO)	77
forests	7,29
fossil fuels	7,9,85
Gasbuggy programme	114
Geddes, Patrick	119
Great Barrier Reef	6,49-50
Hamilton Trader	58
Hiroshima	102
hybrid sterility	77
hydraulic fracturing	109
hydrogen sulphide	16,32

Inter-governmental Maritime Consultative Organisation (IMCO)	55
International Atomic Energy Authority (IAEA)	76
International Biological Programme	5,139
invertebrates	67,69
Lakes, Apopka	15
Clear	73
Baikal	17
Erie	12-5
Great	13,15,25,72,158
Michigan	13-4,71
Superior	13
Trumman	16
lead	90
Long Island	35
Los Angeles	88
Luangwa valley	4
lemmings	117
malaria	75,84
Manhattan	52
Mellanby, Dr Kenneth *Pesticides and Pollution*	21
mercury	34
Minamata Bay	34
Mumford, Lewis *City in History*	119
National Parks	138-9
Everglades	138-9
Kruger	130
Serengeti	128
Tsavo	130
nitrates	12-3,22
nitrites	22
nitrogen cycle	2,5
fixation	5
North-West Passage	52
noise	121

INDEX

nuclear explosions	112
power	102
power stations	29,107,110
nutrients	9,12,22-3
oil slicks	55-6,59-60
organochlorines	66-70,79
organophosphorous	79
otter	135
oxygen	2,8
pesticides	13,62-84
(and then pesticide chapter page numbers)	
Panama Canal	113,147-8
phosphates	12,22,25-6
photo-chemical smog	88
photosynthesis	3,63
plutonium	111
particulates	87
polychlorinated biphenyls (PCBs)	39-40,74
radioactive wastes	39,107-10
rhinoceros	131
rivers,	
Guyahoga	15,25
Nile	133
Rhine	18-9
Seine	18
Tame	24
Thames	20
Trent	24-5
Ural	18
Volga	18,60
robin	68
salinization	145
salmon	41-2
San Francisco Bay	33
Santa Barbara	46-8,53,60
seals	136-7

sea-otter	135
sewage	12,21-4
plants	22,153
smog	94
social stress	117-8
sodium nitrilo-tri-acetate	26
Southend	56-7
sterile male	66-7
Sternglass, Prof Ernest	104-5
Strontium-90	103,105,110
Sturgeon	17
sulphur dioxide	94-100
systemic fungicides	80
terns, Sandwich	35
TDE	67
Torrey Canyon	44-5,59,60
turtles	137
Unesco Courier	33
whales	40-1
White Fish Authority	37
wolves	134
World Health Organisation (WHO)	77
World Widlife Fund (WWF)	132,135